Fritz Reusswig, Lutz Meyer-Ohlendorf

Social Representation of Climate Change
A case study from Hyderabad (India)

Emerging megacities
Dicussion Papers
Edited by Konrad Hagedorn, Christine Werthmann, Dimitrios Zikos, Ramesh Chennamaneni

Humboldt-Universität zu Berlin
Department of Agricultural Economics
Division of Resource Economics
Philippstr. 13, House 12
10115 Berlin

Tel.: +49 (0)30 2093 6305
Fax: +49 (0)30 2093 6497
www.agrar.hu-berlin.de/struktur/institute/wisola/fg/ress
www.sustainable-hyderabad.de

Contact: emerging.megacities@hu-berlin.de

The emerging megacities discussion papers are available at:
www.eh-verlag.de

ISSN print edition 2193-6927

Emerging megacities Discussion Papers are prepared by researchers working on topics in the realm of sustainable development in Megacities of Tomorrow, a research priority by the German Ministry of Education and Research (BMBF). The papers have been peer-reviewed by a board of external reviewers.
Views and opinions expressed do not necessarily represent those of the Division of Resource Economics.
Comments are highly welcome and should be sent directly to the authors.
We welcome contributions on any topics related to Megacities of Tomorrow. Further information on the submission procedure is given at:
www.sustainable-hyderabad.de/emerging-megacities

Reusswig, Fritz; Meyer-Ohlendorf, Lutz

Social Representation of Climate Change
A case study from Hyderabad (India)

Emerging megacities Discussion Papers, Volume 4/2010

ISBN/EAN: 978-3-86741-821-8

First published in 2012 by Europaeischer Hochschulverlag GmbH & Co KG, Bremen, Germany.

© Europaeischer Hochschulverlag GmbH & Co KG, Fahrenheitstr. 1, D-28359 Bremen (www.eh-verlag.de). All rights reserved.

Cover: Photo "Metropolis", ferendus (flickr). Creative Commons License

No part of this publication may be reproduced or transmitted, in any form or by any means, electronic, mechanical, photocopying, recording or otherwise, or stored in any retrieval system of nay nature, without the written permission of the copyright holder and the publisher, application for which shall be made to the publisher.

Social Representation of Climate Change

A case study from Hyderabad (India)

Fritz Reusswig[*,†] *Lutz Meyer-Ohlendorf* [†]

September 2010

Abstract

In this report we underline the importance of studying the social representation of climate change for climate policy, especially in a democracy such as India. Social representations are, from a social science point of view, no epiphenomena of 'real' issues, but the very fabric of individual reality and, building on that, collective decision making. If climate change is not socially represented, it is not there in a society. We briefly characterise the Indian climate discourse, which we perceive as being more complex (heterogeneous) than the European or American one. After a brief look at other studies of climate change perceptions, we turn to our own small sample of qualitative interviews (n=16) in Hyderabad, covering a broad range of issues. We then focus on the way our respondents do represent climate change in the context of weather changes, of its causes, and of possible solutions. We present a typology of cognitive maps of climate change, and relate them to the lifestyle and the social context of the respondents that adhere to it. We also try to identify some starting points for a meaningful climate change discourse in Hyderabad, aiming at the improvement of both local adaptation and local mitigation. The report ends with some general conclusions.

Key words: *social representation, perception of climate change, climate change discourse, mitigation, lifestyle, Hyderabad, India*

---

[*] Corresponding author. Tel.: +49 331 288 2576. Email: fritz@pik-potsdam.de
[†] Potsdam Institute for Climate Impact Research (PIK), Research Domain Transdisciplinary Concepts and Methods, P.O. Box 60 12 03, 14412 Potsdam

1 Introduction

1.1 Purpose and scope of the study

Climate change is one of the most striking problems in the world today, not only politically, but also economically. It has gained international political importance, however only in the last two decades. Before, it has long been mainly a scientific problem. This is not only because of its complexity, but also due to its sometimes highly contested character (O'Neill and Hulme 2009; Walker and King 2008). As a consequence, the topic only slowly made its way into the public sphere. On top of that, partly due to its highly scientific character (presupposing large computers and complicated computer models), partly due to their historical responsibility in terms of causing it, climate change as an issue of scientific and public concern has for long been confined to the industrialised world. Due to their growing responsibility and an emerging globalised community of a concerned public (Beck 2010a, b), developing countries and their publics more and more engage in the debate about global warming, its consequences, and the ways to deal with it (BBC World 2010).

However, in the Western context, climate change often remains an abstract concept, an issue that is "remote both in space and time; it is perceived as affecting other communities and future generations" (Lorenzoni and Pidgeon 2006). Contrastingly, in India and in many other developing countries, weather extremes already now have a significant and clearly visible detrimental effect on peoples' lives. Climate change will most probably exacerbate existing weather hazards, and it will add new, and in some cases unknown risks (IPCC 2007). As for India, significant changes are projected in temperature, precipitation and extreme rainfall, droughts, river and inland flooding, storms/storm, surges/coastal flooding, sea-level rise and environmental health risks (Revi 2008).

Climate change has become a political issue in India, both at the level of international and of national politics. As a consequence, it has also attracted more and more the attention of the mass media, especially the print media. Our analysis of two Indian daily newspapers (*The Hindu, The Times of India*) has revealed that both the quantity and the quality of mass media coverage of climate change is absolutely at par with comparable newspapers in Europe (cf. Reusswig et al. 2009). But does this response really lead to a better knowledge and understanding among the general public–especially in such a big, socially and culturally heterogeneous country like India? Unfortunately, relatively little is known about how climate change is currently perceived and understood by lay people.

Particularly in developing countries, in the context of rapid transformation processes, citizens' understanding of the implications for their lives will be essential in order to be able to respond effectively. Equipped with the knowledge about the changing climate, people will be able to debate the issues with their families, communities and governments, and discuss the risks and possible courses of action (BBC World 2010). This holds for the two major societal responses to (anthropogenic) climate change: mitigation against the causes of GHG emissions on the one hand, and adapting to the effects of unavoidable climate change on the other. An adequate understanding of the phenomenon seems to be a necessary prerequisite for both mitigative and adaptive action, regardless whether on an individual or more political level.[1] The purpose of this study therefore is to develop a comprehensive understanding of lay people's perception of climate change in the city of Hyderabad, the capital of the state of Andhra Pradesh in Southern India:

- How do people perceive their immediate environment and the weather conditions and how do they reflect on these conditions?
- Do people perceive changes in the weather pattern and how do they explain and respond to these changes?
- What do citizens of Hyderabad know about global climate change and in how far do they understand this complex and complicated issue?
- What solutions do people consider and which role do they play themselves in these solutions?
- How is the social situation and lifestyle of people related to their perception of climate change, and how does this influence the adaptive or mitigative actions people can imagine to deal with it?

Even within developed countries, there are significant social differences in the perception of climate change, and it is essential to understand these differences, because "how 'danger' is interpreted will ultimately affect which actions are taken" (Lorenzoni and Pidgeon 2006). These discrepancies become even more marked when we widen the

---

[1] Adequate understanding is admittedly a rather vague concept, relying on standards defined by climate science. But it is also clear that the goal of 'scientific literacy' does not mean that all lay persons should become scientists, or be able to read and understand scientific publications right away. It aims at a robust understanding of the basic mechanisms of global warming, together with an ability to assess its actual and future relevance, and to be able to think of some basic strategies of adaptation and mitigation. For an example of what such a basic understanding could (probably: at max) encompass see Rahmstorf and Schellnhuber 2007 and Walker and King 2008.

scope of inquiry and take developing countries into account, given the sharper social contrasts that influence the social representation of climate change.

If climate policy in an encompassing sense of the word is to be effective, it needs the active support and participation of everyday actors across the globe. If people lack to understand that global warming is happening, that it will aggravate in the future, and that they individually as well as collectively can do something about it, climate policy as a purely governmental issue will most probably fail. This is the ultimate reason why the study of social representations of climate change is not only a meaningful, but even a necessary endeavour. For the same reason, we need to understand the differences between individual representations, as this will help to address the problems more effectively, viz. to identify group-specific mitigation options and improve the knowledge of the people and therewith enhance their adaptive capacity.

1.2 Social representation of climate change

For the natural sciences who historically detected it, climate change is a complex, but given reality 'out there', consisting of elements such as solar radiation, atmospheric chemistry, and bio-geochemical cycles (Rahmstorf and Schellnhuber 2007; Walker and King 2008). Nevertheless, the long and still not terminated history of the scientific discovery of anthropogenic climate change (Fleming 1998; Weart 2003) shows that it was not easy to reconstruct the mechanism behind it. And the comparison between the rather slow pace of discovery in the first period (roughly from the late 19th century to about the 1970s) with the second, more systematic period since then, where most of our actual knowledge about the climate system has been gained, reveals that science was successful only once the recognition of climate change as a social problem had been established, and had led, among other things, to a substantial increase in resources dedicated to look at the issue (Corfee-Morlot/ Maslin/ Burgess 2007). The formation of the Intergovernmental Panel on Climate Change (IPCC) in 1988 illustrates this qualitative shift.

Despite its scientific core tasks, IPCC is a 'boundary organisation' between science and policy. Its success in terms of putting scientific findings at the fore of political decision making (Skodvin 2000) once more illustrates the importance of science for the public perception of (social) problems.

Climate change is a highly scientifically mediated issue. Other than, say, air or water pollution, average everyday actors can hardly detect changes in the central parameters of the Earth's climate (Yearley 1994). One of the reasons for this is the 'ontological' distinction between weather and climate: While 'weather' refers to the concrete state of selected parameters of the lower atmosphere, such as air temperature, humidity or sunshine, 'climate' is a theoretically more ambitious construct, referring to statistically significant patterns of weather over time, linked to basic mechanisms of the physical Earth system (such as the solar constant or the global carbon or water cycles). This is not to say that lay persons are unable to notice changing climate patterns. People with high stakes in the economic use of climate sensitive natural resources, such as farmers or fishermen, have developed their own methods of monitoring weather patterns in order to cope with adverse effects, especially in developing countries (cf. Broad and Orlove 2007; Orlove 2005; Patt 2001; Semenza et al. 2008). Nevertheless, global climate changes can only be detected by systematic instrumental records, statistical methods, the analysis of historical data sets, and computer models (Edwards 2001, Rahmstorf and Schellnhuber 2007), which are beyond the scope of non-scientific observers. This is why the history of the climate discourse (see below for a definition) is for long almost exclusively—and even today to a substantial degree—a history of climate science (or its predecessors) (Fleming 1998; Weart 2003). Measured global warming until today adds up to no more than an increase of $0.8°C$ of Global Mean Temperature (GMT) since the 19th century (IPCC 2007). GMT is a statistical construct that integrates across all geographical regions, seasonal differences as well as day/night-differences—the very fabric of everyday experience. Usually, perceived and relevant daily temperature changes by far exceed this figure. The scientifically mediated character of climate change especially holds when it comes to the attribution problem: who or what is responsible? Climate (other than weather) refers to long-term patterns and processes of the atmosphere as embedded into other bio-geochemical cycles, influenced by the oceans, the biosphere, human activities, and natural factors, such as volcanic eruptions. Causal analysis and attribution in such complex and non-linear systems is extremely difficult—one of the reasons for various uncertainties in climate science statements, and a major driver behind the increasingly interdisciplinary character of climate science.

One of the foci of the social sciences with respect to climate change is the social representation of climate. There are two terms that need explanation here: 'representation' and 'social'. Let us start with the first one.

It has been mentioned that for both scientific and everyday observers, objects tend to be 'simply there', i.e. something given, a positive fact of our experience, part of the world 'out there'.[2] Closer reflection and inspection will however reveal rather rapidly that this is by no means the case and things are much more complicated. Be it 'weather' or 'climate': any 'given' object is given to an observer (to 'us'). If we stick to the reality of 'givenness' we have to talk about subjective elements: our sensations of temperature or rain, the subjective perspectives and the situations in which concrete experiences have been made, the conceptual and/or language elements by which we formulate what we have experienced, the standards of communication by which we exchange with others in order to check whether or not our experiences are shared ones and thus 'substantial' or 'reliable', and so forth. Far from being simply a secondary phenomenon, the 'representation' of an object (weather or climate) seems to be not only the first thing we can get hold of, it seems also to be the only thing we can seriously grasp. Every claim that, for example, 'out there' we can find something else (e.g. a different temperature than the one 'we' have perceived) has to deal with the fact that we cannot really transcend the realm of our subjective representation. It is not possible to take climate 'itself' and compare it with our representations. We will always end up by comparing representation A with representation B—together with the claim that representation A is a better, more adequate representation of reality than representation B. Science is about the inter-subjective justification of such claims, not about immediate transcending of representations. This is why science stresses the methods by which a certain statement has been achieved, and cannot directly 'let the object speak for itself'.[3]

While this immediately raises the very interesting question if and how such a claim can be justified, it is worth noting that the representation of climate is a very basic notion, and that both science and everyday life rely on it. It is neither easy nor strategically wise (for reasons to be explained later) to dismiss the representation of climate as something irrelevant or inferior.

This brings us to the next element of our term: 'social'. It might seem as if this term would add a characteristic to the term representation, which the latter does not have otherwise. And this is supported by our first intuition, which clearly distinguishes between individual representations and those we might share with others, the latter would then be termed social representations. However, things are more complicated here as well. For methodological individualists, there are no social representations.

---

[2] For the moment it does not really matter that—in our stylised way of thinking—lay people take 'weather' as this object, while scientists refer to 'climate' as a scientifically mediated one. Both would not hesitate to take it as a given object, as a reality beyond the subjective level of representation.

[3] Scientists might use this latter expression in order to qualify a certain result (claim) as superior to another one. But in fact they refer to a better method, i.e. a better way of a community of scientists to come to a particular result.

More precisely: the term 'social representation' is nothing but a *façon de parler*, lumping together a number of individual representations of something, the latter being the only valid case of representations whatsoever.

As we do not have the time here to justify our position, we simply have to state that we do not believe in methodological individualism. Together with an otherwise highly heterogeneous theory coalition of Aristotle, Hegel, Marx and American Pragmatism we take human beings as social beings, and together with Wittgenstein we deny the possibility of a strictly private language, i.e. a language that only an individual can develop and understand. If language is only possible as a shared language (or as a social phenomenon), and if language is constitutive for representation (another claim we cannot justify here), then every representation is a social representation. This does not preclude that there are individual representations, and it neither precludes that the individual experience is really and irreducibly given to the individual. But the term 'social' does by no means exclude this individuality, referring not only to something that people share as opposed to what they might think 'privately'. The "social" is in itself structured as the space of shared individual representations. Individuality, exclusion and privacy are included in the social—the same way that the reference to the other(s) is constitutively inscribed to 'my own' representations. It is thus not difficult to see that we argue in favour of a dialectical conception of representation, with 'me' and 'us', or 'individual' and 'social' being the two contrarian (if not contradicting) elements that at the same time cannot be separated from, but refer to each other.[4]

The social representation of climate change thus seems to be a very important scientific object. Other than via the various ways of influencing the shared or contested views of people, the Earth's climate cannot 'intervene' in society. Even if we assume climate to be a core boundary condition for human action (e.g. for agriculture as a basic achievement in human history, cf. Ponting 1993), humans need to take climate and climate changes into account, i.e. climate needs to be represented in order to influence human adaptation to it. One might argue: "Yes, but a flash flood might simply wash away my house, whether or not I am able to represent it". But this is not exactly true. In slightly modifying an argument put forward by cultural anthropologist Marshall Sahlins (1976) one can state: Water can destroy a wooden construction, but it cannot destroy 'a house'. For the water coming down a river the term 'my house' is meaningless. It operates as a physical agent, exerting physical power to some wooden (or other) materials 'my house'

---

[4] Mutual implication of contrarian terms is one of the major drivers of dialectical thinking in the social sciences; others are negative self-reference and pragmatic presuppositions (Müller 2009).

is built up with. But what makes these wooden constructions 'my house' are something the water cannot destroy: my (shared) concepts and feelings of 'home' (and, possibly, 'family', 'childhood', or 'work', or 'leisure') attached to an ensemble of things. Now why then would people weep in case of a flood destroying their homes, if not for the reason that it was destroying their notion of home together with the ensemble of materials that made it up—and that seems to be the material basis, if not the 'incarnation' of my notion?

To the degree that notions have to 'materialise', floods can destroy homes, not only bricks and walls. But as people usually try to rebuild a home after a flood, there is clearly a core element of 'home' that the flood was not able to destroy. Otherwise people would in the next period built some construction, but by no means a home, which would have been washed away together with the bricks and wood of the old (and obviously only) one.

Our point is: If people do not interpret their built environment as 'home' (and live up to that interpretation), it is no home for them. And whatever the flood then destroys: it is not their 'home'. It might, for example, be their 'intermediate shelter' or even their 'prison', but not a home to which they attach so much emotion.

But the power of social representation goes beyond the level of what is at risk when it comes to climate change. It also relates to climate change itself. Or, to stay with our little example, it refers to the exact nature and meaning of the flood. Whether or not a particular flood (as a single event) can be attributed to climate change is difficult to say, not only for lay people, but also for climate scientists. But other than climate scientists, lay people might not be acquainted with the concept of '(anthropogenic) climate change' at all. As one cannot perceive it with direct (sensual) experience, climate change is present in society mainly through the mass media, which communicate scientific findings as well as political decisions related to climate (Boyce and Lewis 2009).

Humans had to cope with different climate zones and weather extremes ever since they exist. And humans constantly did have to represent what was 'out there', otherwise it would have been 'nothing' for them. Humans have, according to their cultural environment, ever since tried to make sense out of their 'physical' environment. This environment has been represented as a realm of mythological and religious forces long before it had been 'demystified' to a purely 'physical' world in the process of European enlightenment. Adorno and Horkheimer (2002) point out that this 'mythological' or 'religious' period of human representation must be conceived not only as the first step towards enlightenment, but also as a first step towards the creation of a meaningful environment—an environment that humans could live in instead of something that hominids did simply populate.

Nevertheless, the term 'social representation' is a rather recent theoretical concept. In an early paper on social representation, French social psychologist Serge Moscovici distanced himself from a purely behaviouralist point of view:

> ...*it is obvious that the study of social representations must go beyond such a [behaviouralist] view, and must do so for a specific reason. It considers man insofar as he tries to know and to understand the things that surround him, and tries to solve the commonplace enigmas of his own birth, his bodily existence, his humiliations, of the sky above him, of the states of mind of his neighbours and of the powers that dominate him: enigmas that occupy and pre-occupy him from the cradle, and of which he never ceases to speak. For him, thoughts and words are real – they are not mere epiphenomena of behaviour.* (Moscovici 1984: 14)

The social representation of climate, we can paraphrase Moscovici, is not an epiphenomenon of climate as a (physical) reality, it is a reality *sui generis*, and it is the *only* one people can adapt to or mitigate against. And here again the mass media do play an important role, not only by representing climate change as a scientifically detected reality out there, but also by framing it in particular ways, e.g. by identifying culprits and victims, strategies, responsibilities etc.

Medially conveyed images must be understood as cognitive constructs that help individuals to easily reflect on complex issues and to further construct coherent narratives built on single known information elements and rags of knowledge (Jovchelovitch 2001; Moscovici 1988). Such individual ideas, images, thoughts and knowledge are not stable, they are subject of a continuous reflection process that is not confined to the individual level, but carried into the daily social interactions of people. In this way, these images emerge as socially shared, subjective, and interpretive cognitive constructs that aim to connect abstract concepts in order "to reproduce the world in a meaningful way" (Moscovici 1984: 17) and therefore shape reality. They "should be seen as a specific way of understanding, and communicating, what we know already" (Moscovici 1984: 17).

The above described construction of knowledge, the building of lay people understanding, occurs in different social contexts and it emerges differently across various social groups. These social representations differ, depending on the social context in which they emerge and are therefore, socially specific. They "must be seen as an 'environment' in relation to the individual or the group (...)" (Moscovici 1984: 23). The concept of social representation was first coined by Serge Moscovici (1961), who advanced Durkheim's concept of 'collective representation'. He defines social representation as follows:

> *Social representations are a network of interacting concepts and images whose contents evolve continuously over time and space. How the network evolves depends on the complexity and speed of communication as well as on the available communication media. And its social characteristics are determined by the interactions between individuals and/or groups, and the effect that they have on each other as a function of the link that binds them.* (Moscovici 1988: 220)

One does not have to confine the theoretical background of social representations to the Durkheim tradition. We would like to also refer to Durkheim's German contemporary Max Weber, who tried to build sociology from the basic concept of social action. As a distinctive feature to mere behaviour, action is oriented towards a subjective goal, and this sphere of goals can be termed 'representation'. Of course representations include many more elements than simply the goals of actions—assumptions about the social and physical world for example, or normative considerations. The sphere of representations is crucial for Weber's theory. Rational action for example, a sub-type of social action, is characterised by the 'calculating' relationship between means and ends. From an action perspective, the whole world can become a mean, and is thus perceived as a reservoir of potential means. But in reality nothing 'is' a mean. Things simply are what they are. They become means only in the context of the life-world of purposeful and social animals—human beings. In other words: being a mean is being part of a particular social representation.

The general relevance of social representation to Weber becomes clearer once we realise how he conceives the relation between 'material interests' and 'ideas' or 'world-views'—a relation that is also relevant to Marxism.

> *Not ideas, but material and ideal interests, directly govern men's conduct. Yet very frequently the 'world images' that have been created by 'ideas' have, like switchmen, determined the tracks along which action has been pushed by the dynamic of interest. 'From what' and 'for what' one wished to be redeemed and, let us not forget, 'could be' redeemed, depended upon one's image of the world.* (Weber 1946: 280)

One could conceive a theory of social action that uses *interests* as key drivers (motivators), and observer-defined *social positions* (e.g. with respect to resource endowments compared to others) in order to explain human behaviour and interaction systems. Many Marxist or 'functionalist' approaches follow that line of thought. Following Weber we can see that this is at least an incomplete view.

Human action, far from being 'mute' behaviour, is inevitably interwoven with the logic of reasoning as set out in language. Other than by *interpreting* what we do—or what we observe others doing—the very fact of *doing* is non-existing for humans. And as a private language is impossible, interpretation is a social process, taking place in a world shared by a multitude of actors.

Social change and/or scientific progress can thus alter the meaning of one and the same action. The climate discourse provides sufficient examples: While 'car driving' was an action (and a concept) that did relate to many discourse orders (such as the economy or law), it did not occur as a global environmental problem until the climate discourse identified CO_2 emissions from cars as a major source of global warming.

The same holds for interests. A major move in the language game 'blaming the climate contrarians' that climate scientists often make is to highlight their funding by large coal, oil and gas companies. This implies that having assets in fossil fuels, inevitably binds the asset holder to particular interests and significantly limits the scope of possible actions. Although this is empirically often true, it is not necessarily true. Even in the empirically true cases fossil fuel asset holders still need to interpret their assets and evaluate them in the social space and its dynamic in order to detect profitable ways of utilising the resource.

But there are also cases where this link does not hold true. Some fossil energy companies have started to seriously invest in renewable energy sources, broadening their economic portfolio, while others have remained reluctant, maintaining their fossil fuel path dependency. These cases are not exceptions from a rule ('actions follow interests, interests follow physical assets'), but arguments of another rule ('actions follow *interpreted* interests, interests follow *evaluated* physical assets'). The important point is that this latter rule holds for both cases—fossil path dependency and portfolio approach—alike. It is not (necessarily) a different physical asset base that drives company A to diversify, while company B stays its course. It is a different *interpretation* of that asset base in the economic and political landscape that leads one company to divert from the path taken by the other. These differences may arise due to a new assessment of profit rates from renewable energy sources, or from expectations about new government regulations, or from changes in the public perception of the corporation etc.

It is important to note that Weber, in his quote from above, does stick to the assertion that *interests* govern our behaviour, not (pure) 'ideas'. But if we have a closer look at the meaning and implications of the term 'interest' we will soon realise that the sphere of ideas shapes and concretises these interests, and thus gears the course of interest-driven action.

More recently, social scientists have engaged more thoroughly into the understanding of climate change and climate-society interactions. Some of that literature can be seen as a continuation of previous work on environmental awareness/consciousness and behaviour (e.g. Ernst 2010; Kuckartz 2010). Others try to incorporate the recently grown public attention to climate change into larger concepts of social change in (reflexive) modern societies (cf. Beck 2010a, 2010b; Giddens 2009). Others again have argued that we are currently witnessing a major change of the social discourse on climate (Egner 2007; Reusswig 2010).

The social representation of climate change focuses on the dimension of individual or collective knowledge (in a wider sense), while the term 'discourse'—inspired mainly by Foucault—highlights the dynamics of argument exchange and narratives in larger social settings, explicitly taking the power or other resource bases of the speakers involved into account.

A Climate Change Discourse (CCD) is a thematically focused and (more or less) coupled sequence of publicly visible arguments in various contexts (or framings) that different social actors are engaged in, in order to influence (1) one another, (2) specific boundary conditions of social action (such as politics), and (3) the general public so, that the resource endowments, interests and world-views of the speaking actors have a higher chance to prevail in the social interpretation and individual or collective decision making processes (cf. Reusswig 2010).

As the social representation of climate change does focus on how single climate issues or causal relations between them are given to an individual and publicly presented, one can regard the social representation as the semantic dimension of a discourse, mainly materialised in individual or collective ideas. A CCD would then be the overarching domain, explicitly including the pragmatic dimension of a language game, especially the aspects of *power* and *interests*. If these latter aspects are included, an analysis of social representation translates into a discourse analysis. While social discourse analysis widens the scope of social representation analysis by explicitly focusing on the concrete aspects of public argument exchange, of power and of interests, social representation analysis can be limited to the reconstruction and evaluation of the arguments and their forms in use.

The current paper does focus on the social representation of climate change in India, and particularly in Hyderabad. Its major basis are qualitative interviews with individuals from various social and cultural backgrounds. We have already looked at the mass media climate discourse in India, based on an analysis of two English speaking national newspapers (Reusswig et al. 2009), and we complement this analysis by an analysis of a selection of Telugu speaking local newspapers (see below). As a third element, we are currently analysing the national and international Indian climate policy, namely the National Action Plan on Climate Change (NAPCC). Taken together, these three pieces give a good overview of the social discourse on climate change in India (and in Hyderabad particularly) at a given point in time (GOI 2008).

What motivated us to ask everyday people about their understanding of climate change was the simple conviction that early U.S. sociologist W.I. Thomas expressed like that in 1928: "If men define situations as real, they are real in their consequences." Translated into the climate situation: "Whether and how people perceive (anthropogenic) climate change is decisive for their future climate behaviour—be it as consumer or as citizen." Any climate policy, whether it aims at adaptation or mitigation, is in need of the active support or at least the acceptance of society—at least of strategically relevant minorities.

Today, we know a lot about the public perception of climate change in the developed world. Besides empirical studies and conceptual reflections, climate issues have made their way into standard as well as irregular surveys (as provided by organisations like Gallup or the Eurobarometer). Much less research has been undertaken to learn about the ways in which people from the developing countries—i.e. the majority of the world population—think about global warming. This parallels the situation that we also have a similar lack of analyses of developing countries' mass media discourse on climate change (see Boyce and Lewis 2009). The actual report wants to help closing this gap—and at the same time to widen our scope with respect to the existing plurality of modes representing climate change.

1.3 The natural science perspective: Climate change and local consequences for Hyderabad[5]

India is most likely to experience climate change impacts significantly, however, it is not an easy task to identify localised impacts on a small geographical scale, such as required for Hyderabad. Different kinds of uncertainties have to be taken into account. First of all, we do not exactly know how the main drivers of climate change—anthropogenic greenhouse gas (GHG) emissions—will evolve over the next say 100 years. The global emissions depend on a multitude of factors, including global population development, global economic development, the degree of global economic integration, technological innovations, consumption choices, and environmental policies.

In order to reduce the complexity involved, the IPCC has launched an expert based scenario process in order to generate a set of emission scenarios, based on a plausible, coherent and systemic structuring of underlying economic and other driver models. As a result, 'families' of scenarios (usually referred to as SRES scenarios, based on the initial Special Report on Emissions Scenarios) have been developed that can be used to generate different climate change and climate impact scenarios, with temperature and precipitation changes being the most important climate variables (Nakićenović and Swart 2000).

To cover a wide range of possible developments, it has been decided to select a high- (A2) and a low emission (B1) scenario. In a second step, an Atmospheric-Oceanic General Circulation Model (AOGCM) has been used to calculate the expected climate change for each of these scenarios. The considered time slices are 1981–2000 (reference climate), 2045–2065 and 2081–2100. We evaluated these model runs to obtain projections of the four most impact-relevant climatic characteristics for Hyderabad (e.g. McMichael et al. 2004; Revi 2008; Satterthwaite et al. 2007):

- frequency distribution of daily precipitation,
- total annual precipitation,
- probability and duration of heat waves, and
- annual mean temperature.

Although the full-fledged uncertainty analysis is to be conducted in the near future, we can already give first ranges for quantitative best-guess projections, resulting in

---

[5] This section relies on the work of our PIK colleagues Matthias Lüdeke and Martin Budde as part of their Sustainable Hyderabad work package (WP 1) on climate change impacts and adaptation.

a much differentiated picture (cf. Figure 1). For daily precipitation greater than 80 mm/day in Hyderabad (medium to very high impact intensity with respect to damages) we calculated for the high emission scenario an increase in frequency of about 70 % ($\pm$6) until 2050. This worsened situation tends then to be relatively stable until 2100. For the low emission scenario the opposite is the case: here we expect a low increase (17 % $\pm$15) until 2050 followed by a period of rapid increase (93 % $\pm$15, compared to 2000). In case this result will be finally corroborated by the full uncertainty analysis, one can conclude that a massive reduction of the global CO_2 emissions (B1 instead of A2) will certainly 'buy' some time for Hyderabad to adapt to more intensive rainfalls (in the second instead of the first half of the 21st century), but will not spare the city to prepare for about a doubling of strong rain events. This is even worsened by the result that the subset of extreme events greater than 160mm/day (like, e.g., in August 2000) will increase over-proportionally. At the moment, we cannot exclude a quadruplicating of the frequency of these mega-events (i.e. a 40 year event will occur each ten years–on average).

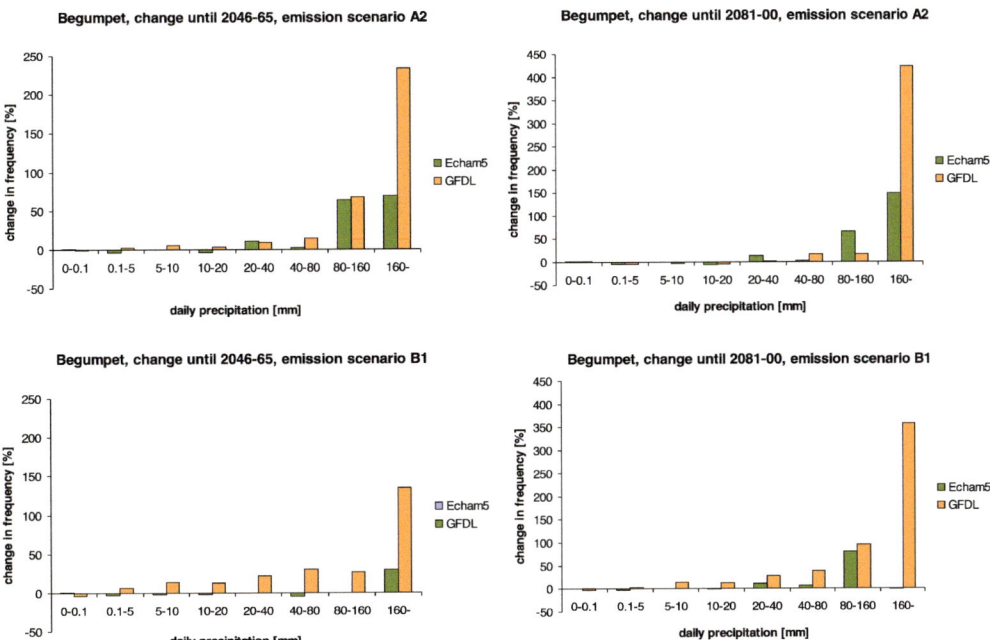

Figure 1: Projected relative changes in the frequency of daily precipitation events at the Begumpet station as calculated from two different AOGCMs and for two different global emission scenarios

Source: Lüdeke and Budde 2009

With respect to heat waves, we first focused on a first indicator, the average number of days per year with night temperatures >27°C. This number will approximately triple until 2050 relatively independent from the emission scenario. In 2100, A2-emissions will lead to an increase of almost 560 % (±50) of the current number while under B1-emissions we expect only a 240 % (±50) increase. The frequency of heat waves longer than one week will double to triple until 2050 and increase further until 2100. The mean annual temperature will develop monotonously in time and with a stronger trend in the high emission scenario up to +5°C. This high value of +5°C for the A2 scenario in 2100 would definitely alter the natural water balance towards increased dryness, even under a (very uncertain) increase in total rainfall. Here a global emission reduction along the B1 scenario would ease the adaptation pressure, but not fully remove it.

It is clear from these results that the city of Hyderabad, i.e. its local government, but also its economy and the citizens, will have to adapt to unavoidable climate change. If no targeted adaptive measures are taken, the city is at risk in various respects:

A probable water shortage can affect industry and services, as well as public supply. Given the competition for water in the region today between industry, agriculture and the residential sector (van Rooijen/ Turral/ Biggs 2005), additional water stress, conflicts and ruptures cannot be excluded.

The strategy of expanding the long-distance water supply as well as the agricultural practice of (subsidised) ground water mining could experience serious limitations, either in terms of general availability (e.g. in the case of region-wide droughts or reduced groundwater recharge), and/or of increased costs. Negative distributional side effects can in that case be expected (Kumar/ Tiwari/ Jha 2009).

Temperature increasing—possibly reinforcing the urban heat island effect—will most probably lead to higher energy demand for cooling purposes, which in turn—given the existing energy structure in India—will contribute to additional GHG emissions (Sivak 2009).

Climate extremes adversely impact transport infrastructure, e.g. by overflow of streets (a frequent problem during extreme precipitation). Heat waves and average temperature increase slowly damage railway and road infrastructure, while making the public transport a very uncomfortable service to use (Shukla et al. 2003).

Impacts of climate change on the health risk can be the direct exposure to heat with the increased occurrence of heat waves and exposure to flooding with a potential increase in serious precipitation events. Flooding can also lead to contamination of freshwater with bacteria, chemicals or other hazardous substances (Young/ Balluz/ Malilay 2004) and its consumption can result in diarrhoeal diseases, cholera and intoxication. The situation is particularly severe in areas of Hyderabad where sewage flows in open ditches close to water distribution pipes (Vairavamoorthy/Gorantiwar/Pathirana 2008) and where people live in industrial areas close to factories (Kovats and Akhtar 2008).

Climate sensitive diseases like Malaria, Dengue and Chikungunya, might increase due to favourable climatic conditions of temperature, humidity and breeding places of water (Bhattacharya et al. 2006) that accelerate the rate of the breeding cycle of the mosquitoes. Intermittent rainfalls, such as in June and July, provide perfect breeding grounds in stagnant urban water like wells, car tires, bottles and cans. In particular, Dengue cases have reached alarming levels in Hyderabad; while the trend of Dengue is going down in the state as a whole, it is going up in the city.

Direct impacts of climate change on food security can be destroyed food stocks or destroyed crops due to flooding or drought. Small peri-urban, but also urban farmers that directly rely on their food for reasons of subsistence or on their income from selling the crops are especially vulnerable (Buechler and Devi 2003; Smith/Garrett/Vardhan 2007). In the short term flooding can also lead to an interruption of food supply to people in some areas.

The question now arises as if and how decision makers and the general public as potentially vulnerable do know about these climate risks, and if and how they see a chance to deal with them. The social representation of climate change and local climate change impacts is crucial in order to facilitate adaptive measures at individual and collective levels. Otherwise these potential impacts might occur, but might find those most affected totally unprepared. Disaster preparedness is one of the most important aspects of disaster loss prevention (cf. Oxfam America 2004; Wisner 2001). Before we address this core issue, we would like to have a short look at the wider picture of climate policy in India.

1.4 Climate Change from an Indian Perspective

The global community of climate and climate impact scientists more or less agrees that climate change is a real concern already now (and especially in the coming decades), as has been very powerfully revealed by the latest report of the Intergovernmental Panel on Climate Change (IPCC) in its Fourth Assessment Report (IPCC 2007). The same community widely agrees upon the general finding that the most severe impacts of climate change will occur in the countries of the developing world, of which India is an important part. A natural interest in both adaptation to climate change and to mitigation policies at a global level thus seems to be a reasonable assumption. For the case of India, a more or less symbolic aspect adds to the former two: The current head of IPCC is an Indian, Prof. Dr. Rajendra Pachauri, and he was—together with former U.S. Vice President Al Gore—awarded with the Nobel peace award in 2007. A more substantial aspect behind this rather symbolic one is the fact that India is home of many first class research institutes that address climate change issues in various aspects; the Delhi-based Energy and Resource Institute (TERI), of which Prof. Pachauri is the director, being only a prominent example.

In late 2009/early 2010, a vivid debate took place, both among scientists (especially, but not exclusively from the IPCC) and in the mass media, criticising IPCC in general and Rajendra Pachauri in particular for some mistakes in the 2007 IPCC report, and for the way the IPCC chair did deal with these mistakes. It seems as if the scientific substance of the failures in IPCC's Fourth Assessment Report is rather limited, so that mass media exaggerations like 'Climategate' are inappropriate (Rahmstorf 2010). However, IPCC and especially its chair Pachauri did a rather disappointing job in dealing with the accusations. Polls in the U.S. reveal a significant detrimental impact of this public criticism of climate science (Leiserowitz/ Maibach/ Roser-Renouf 2010). However, the U.S. public debate on climate change had already been very polarised (in part due to the strategy of the Bush government to downplay climate risks, in part due to the long-standing mass media habit to present a 'balanced'—in fact biased—view of climate science (Boykoff and Boykoff 2004) which only recently had been overcome (Boykoff 2008)[6]. It remains to be seen if and how the Indian climate discourse has been affected, but the empirical data situation is poor. Two respondents in our field study did mention doubts and uncertainties with respect to the scientific basis of anthropogenic climate change, but our own results cannot substitute for large-scale surveys and panels. Nevertheless, they can be a starting point for further research.

Politically, the topic of climate change has gained momentum in India at a very high level, for several reasons:

At *first*, India faces rigid pressure in the international climate change negotiations towards a Post-Kyoto agreement[7] and it retains a tough position against any legally binding commitment. After independence, Indian foreign policy has ever insisted on

---

[6] More marked oscillations are thus no surprise for the U.S. debate, while in Europe mass media coverage of climate change has been much less polarised, and more stable with respect to the attribution problem (Boyce and Lewis 2009).

[7] India contributes almost 5 % to the overall global GHG emissions and the country ranks among the five world largest emitters (World Resources Institute 2010).

maintaining its own sovereignty, especially since the founding of the Non-Aligned Movement in 1961 (Wagner 2010). Economic development and poverty alleviation are among the major priority areas that India is dealing with for more than the last two decades[8]. Against the historical emissions of the industrialised countries, India's very low per capita emissions, and its relatively low carbon intensity (a measure of carbon dioxide emissions per unit of production) (IEA 2006), it is not surprising that India rejects any legally binding international agreement[9] (Wagner 2010).

Second, India is interested in the success of a Post-Kyoto agreement. Its strong position does not mean objecting international climate policy, quite the contrary. Growing environmental problems and the importance of energy security as a precondition for economic development certainly provoke the interest in a successful international climate policy regime – in favour of non-Annex I parties (Wagner 2010). Such an agreement provides the opportunity of assistance primarily in the form of a technology partnership and strategic climate and energy relations with Annex I Parties such as the USA (Bapna 2009).

Third, due to its location, size and population density, India will feel most climate change impacts directly, such as the melting of glaciers in the Himalayans, rising sea levels, weather extremes like droughts and floods, and it might even be facing the destabilisation of the Indian monsoon, with severe consequences for its agriculture and water supply. Given the enormous size and variation of India's physical and social geography, possible impacts, as well as its adaptive capacity will result in a broad range of vulnerability to climate change (O'Brien et al. 2004; Shukla et al. 2003).

Accordingly, also the mass media have taken up the issue, and to a large extent the discourse circulates around the political economy of climate change. This is especially true for the English speaking printed mass media, where the topic is well covered and the discourse stands out by its depth and quality of reporting, as well as by the vividness of debates (Reusswig et al. 2009). An analysis of the local discourse in Hyderabad shows that the Telugu speaking newspapers do cover climate change by far less than the English print media. It was striking also that the newspapers do often refer to weather related problems or disasters, but rarely link them to climate change. However, especially on the local and regional level, it is important with regard to the future development of India

---

[8] India can not be lumped with China as a global economic player with its three hundred million Indians–more than the entire population of the United States–surviving on less than a dollar a day. Four hundred million Indians lack electricity, seeking to switch lights on, not turn them off (Bapna 2009).

[9] However, as a result of the ongoing international pressure India said it is prepared to cut its carbon intensity by 24 % compared with 2005 levels by 2020 (Ramesh 02.12.2009).

and the fate of the poor in particular, to take climate change more seriously, as well as to think thoroughly about adaptation. We have found that many scientists, organisations and government bodies have the same perception, although—from our perspective at least—the process of mainstreaming adaptation to government policies still lacks the necessary vigour and coherence.

More astonishingly though, many actors in India at various regional and political levels do in fact actively engage in climate change *mitigation* activities. At least from an outside observer point of view these activities seem an appropriate answer to the fact that India as a rapidly growing country—both with respect to the number of people it is home to as well as to the size of its economy—not only counts as a major emitter of Greenhouse Gases (GHG) already now, but will do even more so in the near future. Given its rapid economic growth, India will very soon lose one of the major arguments against Western countries: by 2030, the country will bypass Japan in terms of its accumulated emissions, and it will then most probably be the fourth largest emitter of GHG worldwide (Botzen, Gowdy and van den Bergh 2008: 572). With per capita emissions still below Western standards, India as a nation is nevertheless not only subject to climate change generated elsewhere, but at the same time part of the problem.

Our Stakeholder analysis (Reusswig et al. 2009) suggests that even those stakeholders that do *not* accept the latter formulation want to be part of the climate *solution* though (Das, Mukhopadhyay and Pohit 2005). Climate and energy related activities, including various policies and programs at all political levels have been put forward. One driver of this development was the fact that—after some initial scepticism—the Government of India decided to utilise the Clean Development Mechanism (CDM) of the Kyoto Protocol as an important instrument to financially support energy efficiency and emission reduction projects, such as wind farms, biomass or waste based energy generation projects (Parikh and Parikh 2002). By now, the CDM is perceived as a potential instrument for win-win benefits, aimed towards local economic development and environmental improvements concomitant with controlling greenhouse gas emissions.

India is also playing a key role in international policy formation on greenhouse gas emissions. The country has strongly advocated that long-term greenhouse gas emissions should be the same per capita throughout the world—an equal human right to use the global commons.

The fact that the international community (especially the U.S., but also the EU and Japan) has increased its pressure on the Indian Government during the COP 15 negotiations in 2009 to accept internationally binding emission limitation goals can offer only a partial explanation.[10] A more careful look at these programs reveals that most of them link climate issues with other problems of the country—such as rural development, environmental protection in general, energy security, or resource conservation (GOI 2008). If we term those other aspects as elements of sustainable development, one can reasonably argue that many activities of stakeholders that in fact either improve energy efficiency or reduce GHG emissions by alternative energy sources are often at least as much framed under the heading of 'sustainable development' as under the heading of 'climate policy'.[11]

Against this background, we can draw some basic lessons from the Indian policy context for an analysis of social representations of climate change in India: (1) Given the high relevance of global equity considerations in India, and the widely shared view that India's low per capita emissions are a major anchor point for climate policy debates, we do not expect a high pressure on mitigation actions from the side of the political discourse. This is reinforced by the poor economic situation of large majorities. (2) Given the rather elaborated institutionalisation of climate or climate relevant policies (e.g. a ministry for renewable energy, and a National Action Plan), together with quite active NGO and science sectors, reflected and reinforced by a rather good coverage of climate change issues in the English speaking national newspapers, we expect a rather well organised social representation of climate change (and climate policies) at the level of high-end decision makers and higher educated people. (3) Given the low attention that regional language newspapers give to climate change (not to weather extremes!), and given the rather 'expertocratic' mode of the NAPCC introduction so far, we expect a low degree of elaborated representation of climate change at the level of less educated and/or less powerful people.

---

[10] In fact, in the Indian case this pressure might even lead to increased opposition to binding climate goals. Many experts we have been talking to tend to believe that the situation in India is really different to the Chinese case, which for Indians is always a major landmark for orientation. Chinese climate policy initiatives can, at least to a higher degree, be traced back to the very sensitive reaction of leading Chinese (communist) politicians to international pressure, whereas India, as a long-standing democracy is a much more 'accepted' member of the international community, resulting in a somewhat higher self-esteem when it comes to turn down international claims that are perceived to violate Indian interests.

[11] This is a strategic embedding that is by no means limited to the Indian situation, but can be found in many other developing country policies addressing climate change. *The Caribbean Community Climate Change Centre* (CCCCC), for example, located in Belmopan (Belize), uses sustainable development as a 'hook' in order to mainstream climate change issues (adaptation, mitigation) into existing social and political processes in the Caribbean context (Dr. Neville Trotz, CCCCC, personal communication).

2 Review of previous studies on perception of climate change

2.1 The BBC Climate Survey

In the summer of 2007, briefly after the publication of IPCC's Fourth Assessment Report, the BBC World Service commissioned a poll of 22,000 people in 21 countries with respect to various climate change issues (BBC World 2007). The sample covered people from industrialised as well as from developing countries. In India, 1,521 persons participated. Results to four questions are of particular interest here: (1) To what degree people have heard of global warming or climate change, (2) whether or not they perceive human activities as a significant cause of climate change, (3) whether or not they agree on action to address it, and (4) what their position was regarding limiting GHG emissions in less-wealthy countries.

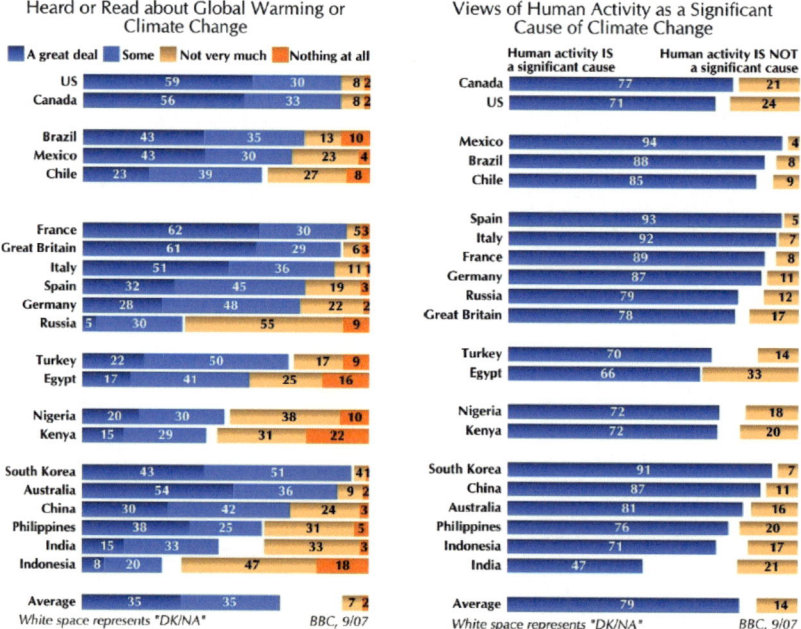

Figure 2: Share of people that have already heard about climate change (left) and share of people that view human activity as a significant cause (right)
Source: BBC World 2007

With regard to question 1 (heard or read about climate change) it strikes that the Indian public is well below the world (more precisely: the poll) average: 70 % have heard or read about climate change on average, while only 48 % in India.[12] In contrast, the share of those who have not heard very much or nothing at all in India is 36 % as opposed to 9 % on average.

This result indicates that both the mass media coverage of climate change (or global warming) as well as its role in everyday communication in summer 2007 was relatively poor in India—at least the public impact of any possibly existing coverage. One might attribute this poor presence of climate change in India's public discourse to the relatively low stage of economic development—a very common argument both in developed and developing countries. While this assumption holds true in many cases (cf. Egypt, Kenya or Indonesia), people from countries with comparable levels of development have heard more of climate change (e.g. China, Philippines, Egypt, Brazil).

With respect to question 2 (humans as significant cause of climate change), much less Indians (47 %) believe that climate change is caused by human activities than the sample average (79 %), whereas slightly more Indians (21 %) than on world average (14 %) believe that human activity is not a significant cause. This low degree of attribution—in fact the lowest in the sample—gives rise to concern about (a) the public understanding of climate change science in India, as well as (b) to the prerequisites for mitigation action.[13] On the other hand, one must not overlook the fact that more people in India agree that human activity is a significant cause of climate change than those believing the opposite.

When asked whether or not action is needed to address climate change, only a slight majority of Indian respondents agreed that major steps soon were needed (37 % as compared to 65 % on average), while 26 % (average 25 %) only wanted to see modest steps in the near future; 12 % (average 6 %) thought that it was not necessary to take

---

[12] If we narrow the answer category to 'a great deal', assuming that only these people make up the core of rather well informed citizens, India's 15 % contrast even more starkly with the average of 35 %.

[13] With respect to the former, it is interesting to compare India with the U.S., a country much further ahead in terms of economic development. Nevertheless, attribution of climate change to human activities is below world average here as well, although not as marked as in India (71 % against 47 %, average being 79 %). While scientific and educational infrastructure in the U.S. is very good, the public perception of climate change has been heavily influenced by a federal government that more or less officially doubted the scientific validity and certainty by which climate change was termed 'anthropogenic', and by a mass media coverage that reinforced sceptical minority positions by a balance-as-bias attitude (Boykoff and Boykoff 2004). Concerns about mitigation action are a logical consequence of the low degree of human attribution: why should one support mitigation policies (or actively engage in GHG emission reduction activities at a personal level) if climate change is a *natural* phenomenon? This latter concern holds independent of the question whether Indians believe *what* particular human activities—and especially: *where located* (inside or outside India)—may or may not contribute to global warming. If human activity in general was no cause of it, then there is no point in holding the industrialised accountable either.

any steps at all.[14] Again, the comparison to China is interesting, as in this country 70 % thought that major steps should soon be taken, and only 4 % (even below world average) assumed no steps at all to be appropriate. Both countries find themselves in a structurally equal position: relatively 'early' stage of development, low per capita GHG emissions, up to now a low historical carbon footprint, large and growing total emissions, and a rejection of internationally binding emission reduction agreements, while at the same time taking some actions to improve energy efficiency and security. But only in China, a clear majority seems to support these policies by their personal attitudes.

A very interesting issue is raised by question 4: "Because total emissions from less wealthy countries are substantial and growing, these countries should limit their emissions of climate changing gases along with wealthy countries." Fifty-nine percent of the total sample hold this view, while 29 % oppose it. In industrialised countries, consent is even more marked (USA: 75 %, UK: 70 %, Germany: 61 %). The Indian respondents are clearly split: 33 % agree that developing countries with substantial and growing emissions (such as India itself) should limit emissions, 29 % oppose this view, and 43 % are not decided. While respondents from the developing world tend to agree slightly less, India's rate of agreement ranks lowest in the total sample. Agreement in Nigeria is 42 %, in Kenya 64 %, and even in China, which finds itself addressed implicitly by the question, is 68 %-the same as in Canada or the UK. If we compare the opposition to this point, i.e. people who think that less-wealthy countries should not be expected to limit their emissions, we find that the Indian value is lower than average. 53 % of Egyptians, 50 % of Nigerians, but also 49 % of Italians or 34 % in Germany hold that view.

We can conclude that the Indian public seems to be rather fragmented when it comes to limiting GHG emissions in India itself. The issue can receive no strong support, nor does it face strong opposition. A majority finds itself undecided. If combined with financial assistance and technology transfer, the situation changes, and almost half of the respondents would support emission caps for countries like India.[15]

2.2 The HSBC Climate Confidence Index

At about the same time as the BBC Climate Survey was published, the international bank HSBC (2007) has published a 'Climate Confidence Index' (CCI), based upon a

---

[14] The question focuses on "steps to reduce the impact of human activities that are thought to cause global warming or climate change" (BBC World 2007), which refers to mitigation activities.

[15] If the emission cap was combined with financial assistance and technology from wealthy countries, 47 % of Indian respondents would agree.

survey among nine countries (n=9,000), including India. While in the UK and the U.S. the survey was more representative, in other countries the sample only covered higher socio-economic groups (in India: urban mass affluent).

The CCI is a composite index, based upon four independent indices: (1) concern about climate change, (2) confidence in what is being done about it today, (3) commitment personally to contribute today, and (4) optimism that we will solve the problem.

In a slight contrast to what we might expect based upon the BBC survey, climate change concern in India is the highest (60 % of respondents say it would be among the biggest issues they worry about) in the sample (UK: 22 %, USA: 32 %, Germany: 26 %, China: 47 %).[16]

Confidence in the people and organisations responsible is medium in India (19 %), with most respondents from other countries being less confident (e.g. UK: 5 %, Germany: 6 %, USA: 13 %, Mexico and Brazil 14 %), and respondents from Hong Kong (38 %) and China (46 %) being much more confident.[17]

If we narrow down the sample to those who rank climate change as being the most important issue for concern, and then ask for what should be done—the HSBC study assumes this to be part of the confidence index—then 40 % of all Indian respondents support the statement 'We should make a big change to all of our lifestyles today to reduce climate change (Make a big change)', while another 40 % support the statement 'If we all act now, we can help stop climate change for very little cost or disruption (Impact with a small change)'. Only 20 % support the statement 'It is not our role to try to interfere with the earth's climate (Don't try).'[18] This latter option—a hodgepodge of

---

[16] If we were to assume that this result is a consequence of the urban middle class bias of the Indian sample we could draw an encouraging lesson from it: other than what some critics of India's myopic middle class consumerism assume, urban middle class Indians seem to be really concerned about climate change. It was outpaced only by concerns about terrorism. Third largest reason for concern in India was global poverty. Concern about climate change increases with age.

[17] We know from other studies that general confidence in government is very high in China (REF). Most probably, Chinese respondents related the phrase 'people and organisations in charge' to their government.

[18] This statement comprises positions that assume a high resilience of the Earth system as well as adaptation or technological optimism ("we will find adaptation/technological fixes to climate change early enough, so do not bother") or fatalism ("we will not be able to solve the problem early enough"). It seems to us that these answers diverge significantly in meaning and should not be mixed together in a single category. In addition, the phrasing 'it is not our role to interfere with the climate system' is misleading, as it resonates with Art. 2 of the UNFCCC, stating the goal of the Convention as preventing dangerous climate change (human interference with the climate system). HSBC's 'it is not our role' does, at least in some of the concrete answers, mean just the opposite: as our 'interference' with the climate system is that strong and related needs or interests that vested, it is hopeless to avoid dangerous climate change.

various attitudes—is much more marked in countries like the UK (33 %), China (32 %), or Germany (41 %).[19]

All respondents across the sample state that the government should take a leading role in doing something about climate change (68 %), followed by individuals (18 %), companies/businesses (10 %), and NGOs (5 %).[20] In fact, the NGOs are taking the lead when it comes to the question who is actually fighting against global warming (42 %), while governments come in only second (33 %); also the business sector is perceived to act less as needed (5 %). Interestingly, besides the NGOs, the individual seems to act also above its level of appropriateness (20 % versus 18 %).

The commitment part of the CCI in India is also highly marked, i.e. Indians more than other respondents from other nations believe that they personally can and do contribute to mitigate against climate change (47 % against 44 % in China, 23 % in the USA or 25 % in Germany).

To a large degree, this high score of India is a result of the fact that Indians—as Mexicans, Brazilians or Chinese—did agree significantly above average to the statement 'I am personally making a significant effort to help reduce climate change through how I live my life today'. Lifestyle changes are seen as the most important individual commitment across the board (58 %), preferred to spending extra time (45 %) or money (28 %).

This is an interesting finding. On the one hand, one can argue that this answer truly reflects the very constrained reality of average Indian lifestyles, resulting in significantly lower per capita emissions of GHG. Would everyone on this planet lead the life of the average Indian, climate change would not have emerged as such a problem.[21] This would require the respondents to be informed about the climate effects of (average) Indian lifestyles as compared to those in other countries. On the other hand, given the urban middle class bias of the sample, averaged Indian lifestyle related emissions are lower than those presumably generated by the respondents. This would indicate that middle class urban Indians identify with the country's average lifestyles and related emissions when it comes to assess their own contribution to climate protection. Should

---

[19] In the UK, the dominant reason behind the 'don't try' option is the conviction of many that we should adapt to unavoidable climate change, while in China a majority believes that scientists will find technological solutions. The by far dominant reason for 'don't try' adherents in Germany is pessimism: it is impossible for us to try to stop it.

[20] Expectations vis-à-vis the business sector in India rank even lower than in the sample average (6 % versus 10 %). The Indian urban middle class seems to hold the government even more responsible than other country samples.

[21] This would, of course, hold even more true if we consider average African per capita emissions. This is why we have termed African consumers 'involuntary climate protectors' (Reusswig/ Gerlinger/ Edenhofer 2004).

their actual emissions be much higher than the Indian average (something we will have to find out during the further course of our research) this might well be perceived as a self-serving defensive mechanism.[22]

With regard to the final component of the CCI, optimism gives least reason for hope: only 23% on average express their assumption that we will be able to stop climate change. In India, however, 45% believe that this can be done—the highest value in the whole sample. This climate optimism of Indian consumers is not easy to interpret. May be it is the result of a significant trust in those responsible to do something, combined with the conviction that individuals already do what they can. In any case, in this sample India is a unique country in the sense that it combines highest concerns with highest optimism that we can solve the climate puzzle.

2.3 The Nielsen Survey

In the weeks before the G8 Summit 2007 in Heiligendamm (Germany), the market research institute "The Nielsen Company and Oxford University's Environmental Change Institute" (Nielsen 2006) released an international comparative study on climate change. The online survey, the largest of its kind to be conducted globally on the topic of climate change was conducted in April 2007 and polled 26,486 internet users across 47 countries in North America, Europe, Asia Pacific (including India), Latin America, Africa and the Middle East.

A major information provided by this survey is related to the dynamic nature of climate change discourse effects. The survey had been done in October 2006 for the first time, and was repeated in April 2007. In every country, the concern about climate change increased substantially during that period—nine percentage points up from 7% to 16%.[23]

In India, concern about global warming increased by 11 percentage points from 8% to 19%.[24] In 2007, India's rate of concerned consumers was thus in the neighbourhood of countries like Germany, Denmark, Mexico or Hong Kong. Respondents all over the

---

[22] Greenpeace India (2007) has termed this mechanism as 'hiding behind the poor'.

[23] Both the Stern Review and IPCC's Fourth Assessment Report have been published in between. The study does not ask for the discourse 'stimuli' that made people more concerned, but we take it to be a valid conclusion that the broad media coverage of these two publications has in fact triggered concern globally.

[24] It is not possible to directly compare the 'concern' results from the Nielsen survey with the 'concern' results from HSBC's CCI. While confined to internet users, the Nielsen survey does also show a bias towards the urban middle classes, but it limits the concern to 'the next six months', which narrows down people's expectation horizon significantly.

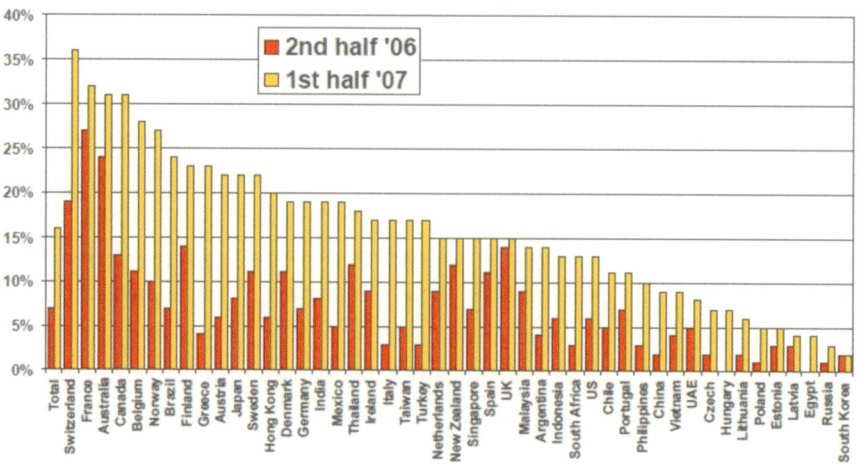

Figure 3: Biggest and 2nd biggest concern in the next six months: global warming
Source: Nielsen 2006

planet on average preferred government action (such as GHG emission restrictions for businesses, incentives or public R&D investment) over individual action, such as recycling consumer waste, driving less car, purchasing seasonal food, home insulation or reducing air travel.

2.4 Africa Talks Climate - The public understanding of climate change in ten countries

Climate change in Africa is not abstract, but already determining the course of the people's lives. Africa's engagement with the issue is evolving rapidly, although most people that are affected are yet poorly informed, particularly outside urban centres.

The research teams held focus-group discussions with more than 1,000 citizens in 10 countries across Africa throughout the years 2008 and 2009 and conducted interviews with 188 opinion leaders, such as policy makers, religious and community leaders, etc. (BBC World 2010). Four questions lead the research teams to investigate their focus of interest: (1) What changes have African citizens experienced in their climate and environment over time? (2) How do African citizens explain and respond to these changes? (3) What do African citizens know and understand about global climate change? (4) What do African opinion leaders know and understand about climate change and what are their views on their country's responses to climate change?

Concerning research question (1) *Africatalksclimate* revealed that Africans have noticed changes in the weather patterns and experience the impacts on their day-to-day lives. Seasons are less distinct, rainfall gets more erratic and intense, and temperature increases. As a result, soils are reported to be washed away or less fertile, rivers to dry up or extend to floods. Consequences of more complex pressures are the reduction of greenery and forest cover and the increase in pollution. The majority of people are concerned about crop failure, loss of livestock, food shortages and illness.

Leading question (2) revealed that the observed and experienced climatic changes were in most cases put in close connection to environmental degradation and natural resource depletion in general, and not seen as distinct from their local interactions with the environment. They were in most cases not related to the changes at the global scale and not seen with particular reference to the developed countries. Much in contrast, many people believe the miserable environmental conditions to be a sign of their own unconscious or unwary behaviour and God's imposition. The biggest concern for Africans, particularly in the rural areas is drought, while deforestation is typically seen as closely linked. Flooding is a great concern in urban areas, often with the blame on lacking proper planning, poor environmental management, unlawful construction and inadequate waste disposal, and the result of overcrowding. These findings were similar across all countries investigated except South Africa, where the processes behind climate change are somewhat better understood. They link emissions from vehicles and industries to the environment and personal health, as well as to climate change, but again much more to their own and the country's development than to the developed countries' emissions on other continents. With respect to their ability to respond, African strategies are reactive and seen as futile when undertaken on the individual level. Consequently, governments are often seen as responsible to step in for both, undertaking abatement actions and providing information for citizens' reactions. Some turn to prayers, whereas the majority of people say that they would be forced to migrate if conditions get worse (interestingly to and from both direction, rural–urban and urban–rural). At the heart of the responses is the belief that individual reactions have little impact on problems caused by groups of people and that the day-to-day struggles are hindering for planed, preventive actions.

With respect to guiding question (3) one of the principal barriers is the terminology. 'Climate' is difficult to translate into many African languages, transport different meaning or refer to the spatial and temporal agglomeration that Africans perceive in more detail. E.g. "in Bari, the only way to say climate change is 'changes in clouds, rainfall, wind, and temperature, and seasons' "(p. 9). In urban areas international translations are more common than in rural areas, but most people find the terms difficult to explain. If people are used to the terminology most of them refer to global phenomena such as cyclones in Asia, heat waves in Europe or melting ice caps. Most of them point to the media or schools when asked for sources of information. During the interviews five important themes that influence Africans' understanding (frames of reference) have been revealed: emphasis on trees ('trees attract rain'), will of god (sometimes independently from human action and sometimes as punishment for humans' misbehaviour–so in parallel accepting humans as cause), ozone confusion (mainly caused by the belief that the 'ozone hole' can let more temperature in), air pollutions (mostly those that can be seen or smelled; smoking cigarettes, bush burning, coking with fire wood, cars and factories are cited as causes of climate change), localised heat (overcrowding as a collection of a lot of body temperature).

Question (4) was addressed to opinion leaders from governments, media, the private sector, religious groups, local and community groups, and NGOs and academic representatives. In those interviews, as well as in the discussions leading to answering question (1) to (3), the exact terms of 'climate change' and 'global warming' were addressed. Knowledge tends to be highest among government and NGO officials, although misconceptions exist on all levels. Local leaders from religious and community groups are least informed and a trusted and respected source of information for the communities. Those opinion leaders that know about climate change share the widespread recognition that the developed countries are to be blamed, that impacts are already felt and that the issue remains an elite debate.

2.5 WWF. Developing an engagement strategy for Earth Hour – India

Having started in 2007 in Australia, the Earth Hour movement has evolved into a large international environmental awareness action, prompting people in multiple cities worldwide to switch off their lights for an hour. The study commissioned by WWF (2009) aimed to comprehend public perception on key environmental issues in each of the selected cities as well as to understand concerns over climate change and motivational attitude to take action. This attitude and perception study is particularly important in a developing country like India, where the majority of the population still get barely 8 hours of energy per day and even the metropolitan cities face the 2-3 hour power cuts almost every day. This makes it hard for the people to understand the relevance and need of another one hour switching off the lights as "earth hour".

As the majority of the respondents belong to the age group of 18-25 years and 26-25 years, it can be concluded that the views and perception of youth has been elicited in the study. To capture mix responses on Earth Hour and other environmental issues respondents having varied educational background were interviewed. Approximately half of the respondent population was at least graduate or above and another half represented all education levels. It can be concluded that respondents were from very diverse groups with occupation ranging from being home managers, working in private sector, had their own business, students to government employees. In the city of Hyderabad, 38 % respondents belong to general caste followed by 36 % OBC and 20 % respondents belong to other castes.

The analysis of the survey results identified that the Hyderabadis were particularly concerned about vehicular pollution as the most important environmental issue followed by air pollution, garbage disposal and management as well as water pollution. Their vision is also to have an eco-friendly city followed by availability of good transport and communication. Other concerns for them were housing and good education facilities. When asked about the responsibilities, 54 % respondents in Hyderabad appeal that everybody should be responsible and should take eco-friendly action. However, 29 % stated that this is the government's responsibility.

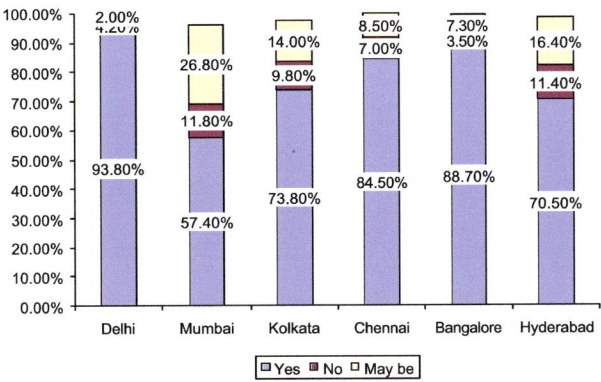

Figure 4: Respondents that believe they live in the period of climate change
Source: WWF 2009

The majority of respondents believe that they are in a period of climate change, i.e. 94 % in Delhi, 89 % in Bangalore, 85 % in Chennai, 74 % in Kolkata, 71 % in Hyderabad and 57 % in Mumbai. Their responses themselves are an indication about their awareness about climate change.

The understanding of the causes of the greenhouse gas effect varies considerably between cities in India, thus reflecting local circumstances. While the majority of people in Delhi highlights air pollution from factories and cars as the main driving force of the greenhouse effect, most of the inhabitants of Hyderabad rather see the cause in deforestation (71 %). The same question when asked in Mumbai did not produce uniform answers, as the causes were relatively equally spread between high urbanisation rate, deforestation and air pollution.

Respondents were interviewed to know if they can relate their city's environmental problems with global warming. Majority of the respondents (88 %) agreed that the environmental problems in their city have contributed to global warming to some extent. 95 % respondents in Delhi, 78 % in Mumbai, 90 % in Kolkata, 93 % in Bangalore, 89 % in Chennai and 85 % in Hyderabad agree to the same.

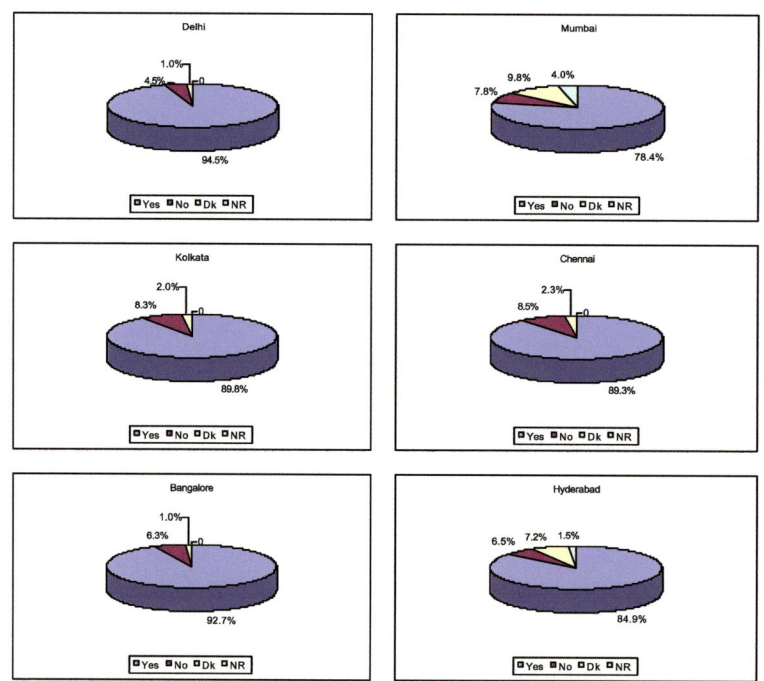

Figure 5: Linkage of environmental problems in the city with global warming
Source: WWF 2009

Earth Hour was not something very well known among India's urban dwellers. On an average only 22 % respondents were aware about Earth Hour. After knowing about Earth Hour, 83 % of Delhiites were willing to switching off lights and 72 % of Mumbaites assured participation. Nearly 85 % in Kolkata, 91 % in Chennai, 90 % in Bangalore and 93 % in Hyderabad respondents will support the cause and participate in Earth Hour. However, some people do not see the need to participate in the Earth Hour because of power cuts and its shortage, unwilling to voluntarily give up something what they are not sufficiently provided with.

3 Methodology

This report is based on a two-month fieldwork conducted in Hyderabad, with the aim to understand peoples' perception of the local climate, their affectedness of strong weather events, and their knowledge and understanding of climate change. Climate change and climate change science are complex, complicated and sometimes contested areas of multidisciplinary research (Walker and King 2008).

In general, lay people develop an understanding of complex issues by constructing coherent narratives with the help of single known information elements and rags of knowledge gathered from media and from daily social interaction between subjects. The product of these social interactions and continuous construction that "shape reality" (Moscovici 1984) are collectively shared, subjective, and interpretive ideas and images, also known as social representations. They "should be seen as a specific way of understanding, and communicating, what we know already" (Moscovici 1984: 17). The creation of these collectively shared images aim to connect abstract concepts in order "to reproduce the world in a meaningful way" (Moscovici 1984: 17). Moreover, "social representations must be seen as an 'environment' in relation to the individual or the group (...)" (Moscovici 1984: 23). That means that these coherent narratives are created in different social contexts and differ, depending on the social 'environment' in which they emerge. Therefore, social representations are socially specific. Lay persons' reflections on complex subjects such as climate change occur through communication, through a communicative process that facilitates the construction of coherent narratives, independent of being scientifically correct or not.

To empirically attend these socially specific ways of understanding and communicating, a flexible and open methodical approach was required. A qualitative research design was chosen in order to fully capture people's perception on climate change, i.e. not only through verifying knowledge of concepts, but through the informants' elaboration on the meaningful interconnections of the used concepts (Flick 2006: 12; Lamnek 2005: 21ff).

The purpose of this study was not only to collect data on peoples' perception of climate change, but also to explore and develop an understanding of the relevance of the lifestyle concept and the character of distinctive behaviour in the Indian context. In addition, the research aimed to enhance the knowledge on social group-specific affectedness from climate related impacts such as heat waves and flood events. For this purpose, 27 semi-structured interviews were conducted, which also contained a quantitative part on personal GHG emissions and basic household information such as household income, consumer durables, migratory background, and cast (see also Table 1).

Table 1: Research design and key aspects

Thematic Focus	Key Objectives	Tools	Key Aspects
Lifestyle segmentation	Understand the relevance of the lifestyle concept and the character of distinctive behaviour in the Indian context based on social position, value orientation and conduct of life	Semi-structured Interviews Household Location Analysis (GPS-Logging, Mapping, Photographic Documentation, Observation Protocol) Informal Talks Participant Observation Newspaper Screening	Value orientation (attitudes, world view, aims in life) Conduct of life (social practice, behavioural and consumption patterns, endowment) Social position
	Identify segmentation indicators for planned quantitative survey		
Social Representation of Climate Change	Understand group specific perception of climate change	Semi-structured Interviews Informal Talks Participant Observation Newspaper Screening	Perception of environmental pollution Energy saving behaviour Perception of "weather changes" Reasoning of "weather changes" Knowledge and ideas of the concept of climate change, reasoning, emotional responses, perceived need for change Solutions with respect to climate change mitigation and adaptation
Climate Affectedness	Understand differing levels of affectedness from climate related impacts		Affectedness from and coping with heat waves, strong rain events and flooding
Carbon Footprinting	First assessment of group specific Carbon Footprint	Semi-structured Interviews	Household electricity use Inner-City Mobility Long-distance Mobility Cooking fuel Meat consumption Consumer Durables

The respondents were selected through theoretical sampling (Flick 2006; Strauss 1987) based on following aspects:

- respondents gender, age, job, income, education, household size,
- housing type and legal character,
- dominant income group of the area, and
- area's distance to the core city

The location aspects were included in the sampling process in order to cover various localities and to understand location preferences of different social groups (e.g. Old City, industrial areas, Gated Communities, Cyberabad, Jubilee Hills, etc.). The location analysis was done through GPS-Logging of the household and GPS-tagged photographic documentation of the area around the interviewed household[25]. Assumingly, a good coverage of the different locality types and different social economic groups were achieved.

A translator was present during most of the interviews, except in those where respondents were fluent English speakers. The interviews were audio recorded and additionally recorded in a rough protocol. The audio data was processed through Verbal Transcription by the researcher in support of a professional Transcriber. The actual analysis was then carried out based on Qualitative Content Analysis (Mayring 2002).

4 Results

The following Table 2 gives an overview of the sample and some social and socio-ecological characteristics of the respondents. Due to some problems in finding professional transcription support, only a subset of the whole data (16 interviews out of 26) has been transcribed for this report. Data from the remaining 12 interviews will be integrated into this analysis in the near future. Through theoretical sampling we have tried to cover the whole spectrum of the local society in Hyderabad, but the small number of interviews does prevent our sample from being representative for the urban society of Hyderabad.[26]

The interview was conducted and communicated to the respondents in a way that climate and environment related issues did not take a lot of space and time compared to other aspects (such as questions regarding quality of life, value orientation, questions on behavioural issues etc.), avoiding a bias in the sense of an expected social norm towards

---

[25] In some cases it was not feasible, e.g. when the respondent was not willing to invite to his home and the interview took place in the office or in a café.

[26] During winter 2010/11 we will realise a representative sample based on a standardised and mainly quantitative study of Hyderabad households.

'climate issues' (social desirability).[27] In addition, we tried to develop an improved understanding of the relevance of lifestyle issues and the analytical usefulness of the lifestyle concept in the Indian context.[28]

Table 2: Overview of the sample

Category		No. of respondents	Category		No. of respondents
Education	Illiterates	3	Household Income [a]	< 90,000	2
	Secondary or below	3		90,000 – 200,000	4
	Graduates	4		200,000 – 500,000	5
	Graduates Technical	3		500,000 – 1,000,000	1
	Professionals	3		>1,000,000	4
Age	18-26	3	Gender	Female	7
	27-36	5		Male	9
	37-49	5	Religion	Hindu	12
	50 and above	3		Muslim	2
Emission levels [tons of CO_2equiv. per capita per year]	Below 1 ton	5		Christian	1
	1-3 tons	5		Sikh	1
	3-5 tons	3			
	> 5 tons	3			

[a] Indian Rupees per household, not per capita. We use the income brackets as defined by MGI 2007.

Respondents were asked to report about their experience of major weather extremes, such as strong rain events, floods and heat waves. Subsequently, they were asked whether they think that the weather pattern has changed in the last 20-30 years or not. The term 'climate change' has not been introduced by us first. Only a few respondents did then spontaneously refer to that concept in the context of their perception of weather

---

[27] Such a bias is a common problem in all areas of social research in the sense that all researchers have to face the possibility that respondents not simply reveal their views and norms, but adapt to what they think the researcher wants to hear (Smithson 2000 cited in Barbour and Flick 2009: 34).

[28] This latter aspect is not at the core of the present report. We will come back to it in later ones. Nevertheless, the interviews used in this report did also have the purpose to prepare our household survey. The relevance of lifestyle issues, especially in the context of India's urbanisation process, has recently been highlighted by Mukhopadhyay 2010.

extremes. If they did, we further asked about possible cause-effect chains in order to get hold of the respondents' mental construct of climate change. If they did not refer to climate change in the context of weather extremes, we continued by asking whether the respondents did ever hear of the concepts 'climate change' or 'global warming' (in English or in Telugu, a local language). This form of open enquiry was also followed with respect to the awareness of environmental pollution. Masked as a question on quality of life in the city of Hyderabad, we tried to figure out whether environmental pollution plays an important role for the respondents' perception of their everyday environment (in a wider sense of the word), without being directly asked about it. We further addressed the issue of energy saving behaviour, which makes sense in an Indian context even without linking it explicitly to climate change for reasons of energy costs and/or energy security.

4.1 Energy saving

Energy saving was a topic for most of the respondents, however, in the least cases due to environmental reasons in general and climate change in particular. In some cases, energy saving was reasoned as a measure to save the resources for the future generations. Mostly, however, people were concerned about the rising costs of energy. Some respondents did, due to their better-off economic background, not bother about energy costs and thus did not engage in energy saving behaviour.

Table 3: Energy Saving Behaviour

Category		No. of respondents
Energy saving (n=16)	Switching off appliances when not needed	6
	Knows Energy Efficiency Labeling orients his/her purchasing decisions accordingly	3
	Using Energy Saving Lamps	2
	Does not bother about energy saving	2
	Sleeping outside in order to save cost for electricity	1
	Car-Sharing to save energy costs	1

These results suggest that behavioural and lifestyle changes, which according to many observers are necessary in the face of global warming (Hubacek et al. 2007, Mukhopadhyay 2010, Roy and Pal 2009), do in fact find a 'resonance space' in Hyderabad. Interestingly, energy saving is an issue also for people who have never heard of climate change, while on the other hand it is a none-issue (or at least a low interest issue) for people who do not have to care for energy costs due to their rather privileged economic position. In these latter cases, other than cost arguments will have to be mobilised in order to stimulate behavioural changes.

4.2 The perception of "weather changes" in Hyderabad

One of the major findings of our interviews was that most respondents do in fact perceive changes in the weather pattern in Hyderabad, but that there are significant differences in the character of these perceptions. Out of 16 respondents, 14 raised concerns about various changes in weather patterns.

Table 4: Perceived changes in the weather pattern in different categories

Category		No. of respondents
Concern about some or the other change in the weather pattern (n=16)	Yes	14
	No	2
Changes in winter (n=16)	Winters have become colder	3
	More fog in winters	1
Changes in summer (n=16)	Summer comes earlier	7
	Summers are hotter than previous	10
	Lack of cool breeze	2
Changes in monsoon (n=16)	Fluctuations in monsoon	11
Changes in precipitation (n=16)	Less rain	7
	Increase of strong rain events	3
	Increase of unusual rains	4

The term 'change' is neutral in the first place, but interestingly all respondents did have negative connotations, i.e. they referred to negatively perceived changes in the weather situation, more or less explicitly stating that the previous pattern was 'better', or that the perceived changes pose a risk to the respondent or the city as a whole. To perceive a deviation from a previous weather pattern seems to presuppose a 'normal climate' (or weather pattern) as a background for observation and evaluation. While from a climatologists' point of view, 'climate' can be seen as a statistical construct, presupposing weather data for about 30 years, everyday social actors—in lacking exact and continuous measurement as well as statistical tools—rely on their own individual heuristics and on communicated experience ('rules of thumb' in order to make up a 'normal' weather pattern as a background for their actual weather observations).

We have already mentioned that people who professionally depend on weather—e.g. farmers or tourist managers—may well develop more sophisticated strategies in developing their heuristics and updates (e.g. by continuous observation of weather reports). However, weather conditions do matter also for urban inhabitants, e.g. street vendors, slum inhabitants, or people with office jobs (e.g. for reasons of how to dress adequately). In addition to this more or less 'functionalist' need to be aware of weather issues, there is a communicative one: in most societies, talking about weather is an easy way to start everyday communication both with strangers (a situation more frequent in an urban context) and with people one already knows. The easiness by which weather issues might be raised as rather 'neutral' (e.g. non-embarrassing) starting points for everyday communication on the one hand indicates their relative unimportance with respect to more 'critical' points, such as the actual personal situation, or political views (Harley 2003). Weather issues thus provide an ideal starting point for 'small talk', and especially weather extremes can be used to enter conversations with almost everyone in every context—something that almost no other issue allows for.[29]

Commonly, most people are able to precisely describe the seasonal weather cycle in its pristine ideal form. The reproduction of this "idealised weather calendar" (Vedwan and Rhoades 2001: 112) has to be understood as the social representation of local climatic conditions.[30] We assume that the perception of a deviation from this 'ideal' climate is again based on a socially shared observation and evaluation that is based on frequent social interactions on the same issue, and the weather is a topic that repeatedly occurs in everyday communication.

---

[29] One might think that issues that affect everyone do offer an ideal starting point for conversation, especially with complete strangers. But this is not the case. Sex, love, educational problems, economic, religious or political issues—all of them would offer ample reason for vivid debates with others. However, exactly because they are usually perceived as 'too hot' topics, people mostly avoid them as entry points for conversation (and often as public issues altogether). The weather affects everyone, but at the same time seems to be 'distant' and 'neutral' enough to serve as a starter. Once climate change, together with the lifestyle and policy choices behind it, has been established as a major driver behind weather extremes, this again might change.

[30] Of course the 'normality' (in the double sense of 'frequent in observation' and 'accepted as norm') of a local weather pattern (climate) changes with geographical location. Human societies have constantly adapted to the natural habitats (including climate) they have occupied and utilised, and they also have developed coping strategies with 'normal' variations, including frequently recurring weather extremes. This, by no means, precludes that inhabitants of region A perceive the climate of region B as 'unbearable'—as either 'too hot', 'too cold', 'too humid' etc.

Only two out of 16 respondents did not perceive any weather changes. From those who did, respondents did significantly more perceive weather changes in summer (19 issues raised) than in winter (4 issues raised).[31] The perceived changes in summer indicate a shift to hotter summers that start earlier. This does relate to the fact that many people (11 out of 16) did mention monsoon fluctuations. With respect to precipitation, most people mentioned 'less rain' (7) as an issue, and some mentioned an increase of strong rain events (3) or of unusual rain (4).[32] These perceptions might not be as precise as meteorological observations, but they also are by no means 'irrational' or 'not sound'. Quite the contrary: they resonate quite well with recently observed shifts—and with plausible scenarios for future climate change in Hyderabad (cf. Lüdeke and Budde, 2009).

4.3 Affectedness by strong weather events

Perceived changes in weather patterns are one thing. Whether or not these changes are affecting people's livelihoods is another matter. One can reasonably argue that urban dwellers usually are less dependent on weather conditions as rural dwellers (especially farmers). Nevertheless, it would also be misleading to assume that urbanites can be seen as totally weather independent and thus 'immune' against changes in weather patterns.

Most of our respondents (12; n=16) never felt directly affected by any flood event. But those who had been affected do live in slums. This clearly resonates with the general observation that socially (and, as a consequence, geographically) marginal groups make up the clear majority of those vulnerable to climate change (and other risks).[33]

For none of the respondents, heat was perceived as an extraordinary burden that has an effect on livelihood or health. It is known from another study (Indraganti 2010) that residents in Hyderabad do have a thermal comfort band in the range of 26–32.45°C, with the neutral (i.e. comfortable) temperature at 29.23°C. This is way above the

---

[31] One person could raise several issues.

[32] We did not investigate the 'everyday meteorology' of our respondents, so that we cannot tell whether they have a consistent way of reconciling 'less rain' with 'more extreme rain events'. We know from scientific meteorology that both statements are consistently possible, e.g. if the first one refers to the total annual sum of precipitation, while the second one might refer to, say, daily maxima during summer.

[33] One of the respondents, who had not directly been affected, raised concerns about food price increases due to flooding. This might indicate a rather 'egoistic' viewpoint, as he seems to care more for indirect flood impacts on him, while the direct impacts to the poor are neglected. However, increasing food prices do of course affect the poor first (Sen 1981), offering a starting point for a viewpoint characterised by more solidarity.

indoor temperature standards specified in Indian Codes. It was also found that subjects living in top floor flats had a higher neutral temperature when the available adaptive opportunities were sufficient. This was due to their continuous exposure to a higher thermal regime due to much higher solar exposure.[34]

Those respondents who are not equipped with space conditioning appliances, such as fans, desert cooler, or air condition, feel the highest burden in coping with heat. Most respondents living in slums also mentioned that they have little means to cope with heat and they sleep in the open during summer nights either due to the heat or as a strategy to avoid additional costs of electricity for running the fan during the night.[35]

4.4 Types of social representation of climate change

Weather patterns can change for various reasons. The Earth's climate history is full of evidence for significant climate fluctuations, which of course did lead to changing weather patterns. However, as mentioned, 'climate' is a statistical construct (or a systemic property of the Earth), and long-term climate changes are beyond the scope not only of individual experience but usually also of social memory. Most people have heard of these long-term climate fluctuations only in the context of mass media reporting about climate science.

Thus, linking perceived weather changes to (global) climate change is a rather demanding task, and without mass media reporting and/or some educational background (e.g. as an issue raised in school or university) people will hardly perceive climate change to be the driver of changes in weather patterns.[36] We have been interested in whether or not respondents could—based on either educational backgrounds, or mass media communication or whatever other source of knowledge—link climate change (global warming) to their perceived weather changes. As mentioned, weather changes had been asked for

---

[34] Although slum dwellers have not been in the sample of this field study, one can assume that they too will probably have shifted their thermal comfort band to the higher end of the spectrum mentioned, due to lacking insulation and/or technical equipment for cooling. The more worrisome is our finding that slum dwellers prefer to sleep outdoor during hot nights: they do so despite being used to hot temperatures.

[35] One of these respondents replied to have had Chikungunya two years ago, which is a vector-borne disease comparable with Malaria and Dengue. This raises the question of secondary health effects of heat waves, such as a higher risk to an exposition to vector-borne diseases.

[36] We would like to repeat or remark that 'professional weather observers' (such as farmers) can in fact develop a lay understanding of how weather changes might be indicative to a change in long-term patterns (local climate). The attribution of human activities as ultimate causes (via GHG emissions and atmospheric GHG concentrations) to global warming, which in turn changes a local climate, which then expresses in shifted weather patterns nevertheless lies beyond the horizon of the practical knowledge even of farmers.

before climate change came into play. 14 from 16 respondents did, as has been reported, perceive weather changes in Hyderabad. Out of these 14, nine did make the link between weather changes and climate change. Nevertheless, these nine people did adhere to very different cause-effect chains, or mental models of climate change. Table 5 gives an overview of the issues raised in order to explain climate change.

Table 5: Acquaintance with and causes of climate change

Category		No. of respondents
Perceives changes in the weather pattern and links it to climate change (n=16)	Yes	9
	No	7
Have ever heard of climate change (n=16)	Yes	10
	Never	6
Reasons for perceived changes in weather pattern (n=16)	Cutting of trees (deforestation)	6
	Pollution	6
	Industries	5
	Ozone depletion	4
	Kaliyuga	3
	Construction of buildings	3
	Waste	2
	Use of plastic	2
	Vehicles	2
	Growth in population	1
	GHG emissions	3
	Carbon	4

A first and very interesting finding is that 10 (62 %) out of all respondents have heard about the concept of climate change/ global warming.[37] While the proportion of those acknowledgeable to climate change is higher in industrialised countries (with some variation), a significant majority of our Hyderabad sample is acquainted with the term. Institutional and behavioural change strategies in order to improve the local adaptive capacity as well as to reduce GHG emissions thus do not have to start 'from scratch', but can rather built upon the concept. A significant minority (38 %), however, has never heard of it. If we control for social status and educational background, we can see that lower classes and people with lower educational degrees (secondary or lower) have a higher 'chance' to remain ignorant about the term.

---

[37] We use percentage points for reasons of easy understanding, well aware of the fact that our small sample size does not allow for percentage points from a purely statistical point of view.

Acquaintance with the term 'climate change' does of course not guarantee that people 'correctly' represent the issue expressed by it. If we assume the scientific 'consensus', as expressed by IPCC (2007), as the 'correct' understanding of climate change, we have to concede that only three respondents (19 %) were able to mention climate anthropogenic GHG emissions as major drivers of climate change. Four of the respondents mentioned carbon emissions as a result of burning fossil fuels as one reason for climate change (all graduates).[38] The professionals (Master or Diploma) mentioned deforestation (6), general pollution (6), industry activities (5), and the depletion of the ozone layer (4) as reasons for climate change. Again, we did not want to correct the mental representations of our respondents, which led us to accept their own models without consistency checks and/or corrected views. This does mean that we accepted folk explanations of climate change that climate science does not, e.g. tracing it back to the (excessive or at least significantly growing) use of plastic (2) or the religious explanation that human greed during a particular age of salvation history ('Kaliyuga') is responsible for climate change (3). We simply follow the rule 'what people think is real at least in some respect', e.g. in terms of how to approach them in order to change their views (if necessary).

In the following, we have tried to systematise the verbal statements of our respondents with respect to the causal network that links weather changes, climate change, and drivers of one or the other. We have used a graphical form of presentation in order to make these patterns more vivid and easy to grasp. We reconstructed the social representations of climate change in individual statements by a kind of cognitive map. One might assume that every individual adheres to a specific, non-interchangeable representation of climate change. As elaborated in Section 1.2, this is a misleading expectation, given the genuinely social character of climate change representations. All kinds of social mechanisms do influence the scientifically mediated view of climate change: mass media reports, school or university education, talks with colleagues or neighbours, etc. In accordance with our theoretical assumptions, our empirical findings reveal typical patterns of different forms representing climate change. It remains to be seen whether or not these types do hold if the sample size is increased. For the time being, the following typology can serve as a starting point for further reasoning, not least with respect to how the local climate discourse in Hyderabad can be improved, and what aspects of climate change should be highlighted and communicated in order to stimulate better adaptation or mitigation strategies.

---

[38] As we did not check for consistency, we did not ask whether people who mentioned 'carbon' would count them as 'GHG emissions', or the like.

In order to link the social representation of climate change to the lifestyle background of their authors, we present a list of the respondents that characterises them in terms of our variables. This includes per capita GHG emissions as well as some energy relevant household equipment.[39]

Every case has a number, and this number will be attached to the links of the respective climate change type.[40] We start with representations that did see no or only weak links between weather and climate changes, and then move to explicitly linked types, ranking them by growing complexity.

4.4.1 Type A: No change, no connection

Our first type of social representation is characterised by the combination of no perceived weather changes and never having heard of 'climate change'. It has to be noted that a changing weather *pattern* does not rule out the perception of weather changes at all. Of course Hyderabad is subject to changing weather, and even extreme events do occur occasionally. However, if a respondent does not perceive this usual pattern of changes to be modified, there is no perceived change in weather patterns. One can reasonably argue that this first type is not really a type in the context of social representation of climate change, for the simple reason that 'climate change' does not occur in people's minds. However, to us the neglect of or the ignorance about this rather demanding concept is part of the way societies 'deal' with it. Put in a more paradoxical manner: even non-representation is a form of social representation. And of course the fact that people are not aware of weather changes and not acquainted with climate change is a serious *fait social* one has to deal with.

Two respondents out of 16 did not perceive any changes in the weather pattern in Hyderabad. They also never heard of climate change before.

This first type does not see any problem with both local weather conditions nor with (global) climate change. One might assume that, given the scientific bias of the issue, this kind of 'ignorance' may be common among people with a lower educational back-

---

[39] We calculated the personal GHG emissions based upon questions regarding energy relevant activity patterns (such as commuting, eating habits, household energy use etc.) and the related technologies used (e.g. cars, fans, ovens). The emission factors (e.g. for electricity or car use) have then be taken from an Indian CO_2 calculator (www.no2co2.in). According to our investigations (subject to another report), average annual GHG emissions (CO_2eq) in Hyderabad equals 1.37 tons, slightly higher than the 1.02 tons put forward by the NAPCC (cf. GoI 2008: 14).

[40] In some cases, not all respondents did state exactly the same causal pattern. Often, some respondents have been more detailed with respect to social causes of climate change than others. Nevertheless, if they occur under the same type heading, they subscribe to key characteristics of a particular type and cannot be subsumed to another one.

ground and/or very limited economic capital. This is not the case in our sample. The two representatives of type A do neither lack school education, nor do they live on low incomes. Quite the contrary: household incomes of 132,000 or 300,000 Rupees[41] classify people as 'aspirers' or 'seekers', the two income brackets on top of the 'deprived' stratum. Both individuals have future aspirations; the female is significantly supporting her family by her work as a beautician, and the male individual comes from a well-off family background of retailing, actually engaged in having fun with his pals while expecting a high-income career in the real estate sector (or the like).

Table 6: Social characteristics of type A

Interview Code		10	20
Informants Age		18	27
Sex		Female	Male
GHG Emissions [b]		1.15	1.36
Household income		132,000	300,000
Consumer Durables	Color TV	1	1
	LCD	0	0
	Refrigerator	1	1
	Fans	0	4
	Coolers	1	1
	ACs	0	0
	Cars	0	1
	Motorbikes/ Scooters	1	4
Education		7th class and Beautician Training	Secondary
Religion		Muslim	Sikh

[b] Overall yearly personal GHG emissions in tons CO_2 equivalents.

The lack of problem perception with respect to climate change can in this case not be attributed to a lack of educational background, or a lack of time to care for a 'luxury' issue due to manifest poverty. Rather one might assume a lack of interest, possibly facilitated by a lack of time due to work and family duties (respondent 10), or due

---

[41] 1 Euro is equal to 65 Rupees.

to leisure activities (respondent 20). In these cases, highlighting the vulnerability of urban systems and groups, as well as the (economic) opportunities of adaptation and mitigation in the city seem important. If one could influence some of the opinion leaders in the social networks (especially in case of respondent 20), the issue would become more 'real' to him.

4.4.2 Type B: Weather changes and pollution, disconnected

One respondent (a taxi driver) did perceive weather changes as occurring in Hyderabad, and he also explicitly took note of environmental pollution (especially: air pollution), but both elements have not been connected by him. They simply occur in parallel.

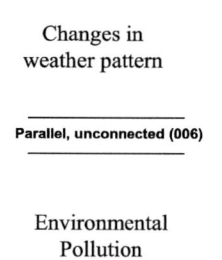

Figure 6: Type B: Weather changes and pollution, disconnected

Due to his job as a taxi driver, our respondent is well aware of air pollution as a core environmental problem of Hyderabad. According to him, lacking regulation is a major cause for this, and if the government would change its policy, cleaner cars would help to reduce air pollution. He does perceive weather changes in the city, but he does not mention nor does he know of climate change. The fact that he perceives pollution as disconnected to weather changes is not surprising from a natural science point of view, as climate change as an intermediary link is not mentioned.

If one would wish to introduce climate change into the worldview of this type of representing things, one would have to highlight the cross-cutting links between reducing air pollution and reducing GHG emissions. And one would have to establish the link between already perceived weather changes and the newly introduced concept of climate change.

Table 7: Social characteristics of type B

Interview Code		6
Informants Age		24
Sex		Male
GHG Emissions		0.78
Household income		120,000
Consumer Durables	Color TV	1
	LCD	0
	Refrigerator	0
	Fans	1
	Coolers	0
	ACs	0
	Cars	0
	Motorbikes / Scooters	0
Education		7th class
Religion		Hindu

4.4.3 Type C: Pollution changes weather

The 'missing linkage' between pollution and weather changes, characterising type B, is overcome in type C: according to respondent 5, pollution directly causes weather changes. Climate change is missing in this person's worldview, as in type B. As indicated, this causal pattern does not hold measured against scientific results. Nevertheless, our respondent does see an anthropogenic driver (pollution) behind weather changes besides climate change.

Surprisingly, despite the incorrect attribution of weather changes to humans, respondent 5 is part of the Indian middle class—a 'seeker' in terms of MGI 2007—and has a high educational degree (Diploma in Mechanical Engineering). He lives in a rather poor area, but works as a computer (hardware) specialist, profiting from Hyderabad's boom-

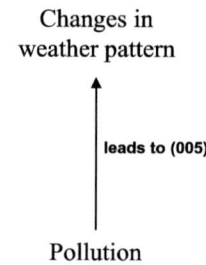

Figure 7: Type C: Pollution changes weather

ing IT branch. The 'task' of a climate communication in case of type B was to create a sound and meaningful link between pollution and weather changes via the introduction of climate change. In case of type C, climate change does also have to be introduced into the mental representation, but at the same time the erroneous idea of pollution as such causing weather changes has to be overcome.

Table 8: Social characteristics of type C

Interview Code		5
Informants Age		30
Sex		Male
GHG Emissions		1.05
Household income		300,000
Consumer Durables	Color TV	1
	LCD	0
	Refrigerator	1
	Fans	2
	Coolers	0
	ACs	0
	Cars	0
	Motorbikes/ Scooters	1
Education		Diploma in Mechanical Engineering
Religion		Hindu

4.4.4 Type D: Religious causation ('Kali Yuga')

Three respondents perceived significant changes in the weather pattern and interpreted them in a religious framework: weather changes as reactions of god to moral degeneration of humans in the dark age of Kali ('Kali Yuga'). As this social representation is rather uncommon in a European context—although ideas about a 'revenge of nature' (sometimes within a religious framework) do also occur here—we will have to elaborate on this type a little more detailed. But first the cognitive map and the social characteristics.

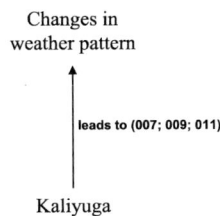

Figure 8: Type D: Religious causation ('Kali Yuga')

Table 9: Social characteristics of type D

Interview Code		007	009	011
Informants Age		45	37	50
Sex		Female	Female	Male
GHG Emissions		0.12	0.35	0.16
Household income		48,000	96,000	36,000
Consumer Durables	Color TV	0	1	0
	LCD	0	0	0
	Number TV Total	0	1	0
	Refrigerator	0	0	0
	Fans	1	2	0
	Coolers	0	0	0
	ACs	0	0	0
	Cars	0	0	0
	Motorbikes/ Scooters	0	0	0
Education		No school education	No school education	No school education
Religion		Hindu	Hindu	Christian

In this group, changes in weather patterns are well perceived, and they are explicitly seen as something negative. They are interpreted as warning signs from god and appear together with other 'signs', basically negative incidents and casualties that cannot be explained easily, such as bomb blasts, flooding, natural disasters, etc.

The social characteristics of the three respondents that adhere to this mental model indicate a very low social status. MGI (2007) categorises annual household incomes below 90,000 rupees as 'deprived', and respondents No.7 and 11 clearly are cases in point; respondent 9 only slightly evades the status of poverty in terms of income classification. At the same time, all three respondents of type D do not have a school education. Their per capita emissions are way below the Hyderabad average, they live in poor areas (slums), and their household equipment is very modest (cf. respondent No. 11). One of the respondents is a Christian (first or second generation), while another one (respondent No. 9, a woman) with rather secure land tenure rights did show sympathies to the Christian religion in our interview.

Nevertheless, these three people did link weather changes to a genuine Hindu notion. Before we can ask why this might be the case, it is necessary to be a little more explicit about the Kali Yuga idea in general. Based on Hindu mythology, Kali Yuga is the fourth and final of an endlessly repeated cycle of epochs (*yugs*) as described in the Vedas. Each *yug* decreases in length and is characterised by an intensifying moral decay (Pinney 1999: 78). According to Hindu traditional calendar, Kali Yuga began on 17th February 3102 B.C., the day when Krishna left the earth to return to his abode. Kali Yuga represents the current age, of which more than 400,000 years remain and "in which moral decay is coupled with natural disaster (...)" (Nelson 2000: 333). One respondent described Kali Yuga as follows:

> *As I know from the mythology, (...) for every 7 villages there remains back only 1 village, the rest will be destroyed, like people die due to floods, earthquakes, and heat. (...) Kaliyuga is nearing; all these signs are the signs for the world to end, like there are so many bombs, over rains, lot of heat. (...) The time comes, when rocks or huge stones fly off in the air due to the wind; all such things are written in that book. (...) It is happening only because of the increase of population and also the sin in the world. (...) Today, (...) I try to talk the truth, I will not be able to feed my family or myself. But if I tell the lie or try to cheat others, only then I am able to earn my daily living; Today, people don't even give water to drink. (...) In the worst days, people will die without food. Right now, (...) it is difficult to eat one meal every day. (...) [People] are trying to kill each other by bomb blast; (...) people die; there is no respect to elders; all these are sins. (...)*

I don't blame God for all these things; we have done it by ourselves and we ourselves are responsible to all the destruction.

Other sequences of the three respondents similarly attribute all kinds of moral misbehaviour (not worshiping god, not respecting the elderly, forgetting about god and moral duties once people become richer, etc.) to the mechanism of purification and eradication of the sinful due to god's revenge.

There is a clear moral bias in this view, not to be found in the former types of social representation.[42] From a Western point of view, the attribution of weather extremes and looming disasters to the revenge of gods for human misbehaviour seems openly irrational and outdated, a lack not only of the scientific understanding of climate change, but of a rational worldview altogether. One might also see this moral/religious framing of climate change as an expression of fatalism, as it attributes it to a pre-determined 'script' of the world, as put down in the holy books.

While this might be an understandable reaction, it would be at least strategically unwise to base public awareness and education campaigns on it. Religious beliefs are not only cognitive world-views, they do also provide social and ethical meaning, and they are often deeply anchored in the affective household of the individual, if not part of his or her personal identity. From this background it seems improbable that an individual would abandon his/her religious belief system simply because some (Western or Indian) climate scientist would explain how climate change 'really' works. In addition to that, one can argue that this 'how it really works' is a less comprehensive and thus incomplete or even wrong description of climate change. Scientific explanations in the Western sense strictly separate observation and analysis on the one hand from moral evaluation on the other. They confine themselves by stating that, for example, car driving contributes to anthropogenic forcing of the atmosphere to X percent, but they would never state 'car driving is often unnecessary and expression of a selfish consumerism', or the like. Nevertheless, these evaluations occur, and everyday actors necessarily evaluate things they observe. From the perspective of the life-world, life is a totality, encompassing facts and values, and acts of evaluation go hand in hand with acts of observation. Evaluation

---

[42] It is not clear from the interviews if and in how far the three respondents using that script do in fact include themselves to their criticism. If we take their poor economic situation, it is hard to believe that they refer to themselves when talking about the growing influence of greed in this world. The same holds with respect to their very modest GHG emission profile. One thus can assume—but not be sure—that these people mean 'the others' (e.g. 'the rich') when talking about the moral disorder of the world.

is a necessary ingredient of observation in the life-world, otherwise it would not make sense to everyday actors.

> *Wie der Sinn der objektiven Welt mit Bezugnahme auf das Existieren von Sachverhalten, so kann der Sinn der sozialen Welt mit Bezugnahme auf das Bestehen von Normen erläutert werden.* (Habermas 1981: 181)

Language and culture are constitutive parts of the life-world (Habermas 1981: 190). Attributing perceived weather (or climate) changes to a morally inappropriate behaviour is not 'irrational' altogether. Only because science has to be mute about the moral judgment of the individual and social driving forces behind greenhouse gas emissions does not mean that such an evaluation is impossible. Even in Western culture, to a large degree coined by scientific rationality, some critical observers attribute environmental degradation as the result of either individual misbehaviour or to the ecological 'blindness' or even the 'perverse' incentive structure of social institutions.[43]

In other words: we should not oversimplify the relation between 'modern' and 'traditional' world-views, not even if they refer to phenomena of the natural world. Jaqueline Homan (2003) examined contrasting social constructions for disaster events in Egypt and England. In one of her case studies that focuses on an earthquake in Egypt, she argues how close the "perception of nature is linked with religious interpretation – for example, concepts of 'oneness' with God and nature" (Homan 2003: 144). Her explanations are analogous with the findings from Hyderabad, in which negatively perceived changes in the weather pattern are attributed to humanities' demoralisation and growing distance from god. Hyderabad as a Megacity represents a good example of an arena of very different and in some instances conflicting cultural perspectives and identities.

Megacities are in many ways "trendsetters and laboratories of the future" and some authors argue that they "have a lot more in common with each other than with their countries or hinterlands" (Nissel 2009). As many other India Megacities, Hyderabad is highly influenced by globalisation processes. Socio-economic polarisation and fragmentation processes can thus lead to "different worlds in one place" (Nissel 2009: 41)—processes that are supported and sometimes reinforced by the still influential remains of the traditional caste system.[44]

---

[43] A lot of the critique of modern consumer culture is characterised by a combination of thorough analysis and of negative (moral) evaluation of individual behaviour and/or social institutions.

[44] In this report we cannot deal with the status, role, dynamics and implications of the Indian caste system (for a comprehensive overview of the classical system see Dumont 2009 and Gupta 2010).

While especially Hyderabad ('Cyberabad') has also profited from globalisation, this complex process does also lead to a widening social gap between the poor and the middle classes, which gets reflected in the educational as well as in the political system of Indian Megacities (Ruet and Lama-Rewal 2009). But not only does 'globalisation' often has contradicting local implications (such as the growth of a well-off middle class with increased demands in consumption and politics on the one hand, and a growing socio-spatial pressure on marginalised groups on the other), it does also lead to contrasting effects in the attitudes and behaviour of people[45].

Religious beliefs and traditions are especially challenged by cultural globalisation, a process that clearly displays a Western bias (Robertson 1992). Despite some 'Christian' attributes that might accompany (cultural) globalisation (e.g. the introduction of Christmas in non-Christian societies), it is basically a non-religious, or faith-indifferent process—if not a profane, anti-religious process, as especially many Muslim critics observe it[46]. We assume that from a religious perspective, the mentioned cultural and social transformations and the direct exposure to cultural influences from the outside world are sometimes perceived as an alienation and estrangement from god. The new way of life of a still small segment of the urban society in India is highly visible (even though not tangible) for those who do not directly benefit from the rapid economic growth of the country. The emergence of more and more shining bureau complexes, shopping malls, fly-overs, gated communities and modern fashion and food cultures are perceived as iconic images for emulation on the one hand, but also as a threat to many people's livelihoods and cultural identity. All this might reinforce the recourse to traditional interpretations.

One might then argue that linking (perceived) climate change to Kali Yuga is inacceptable due to its obviously inherent fatalism: the fourth world age inevitably is an age of moral decay, and there is only one way of redemption, namely the start of a new age by predetermined rules or by acts of God. But one should keep in mind that the

---

[45] While some welcome and embrace the new opportunities in the urban space, others develop strategies of open or concealed resistance. This resistance can take the form of fundamentalist reactions, such as right-wing neo-nationalist movements, ethnic conflicts as in Bosnia and many African countries, or the partly forceful close-down of two Kentucky-Fried-Chicken Restaurants in Delhi and Mumbai. The right-wing movement of the political party Shiv Sena (Army of God Shiva) in Maharashtra since the 1970ies is another example for cultural tensions between local, national and global orientation.

[46] E.g. the introduction of 'Christmas' in non-Christian societies has thus much more fuelled a specific event of mass-consumerism than it has led to filling churches.

interpretation of disaster as signs for a moral decay and as reasons for the revenge of God is also part of the Jewish-Christian and also of Muslim religious tradition.[47]

Within the logic of the Hindu religious tradition, Kali Yuga is inevitable. It is still contended what follows after Kali Yuga, whether it is the beginning of a new and golden age, a 'reboot of the system' that replaces the decay and chaos with a 'divine' order remains unclear in the mythology. However, the normative conclusion from the existence of this predetermined character of the worldly existence is to individually escape from the cycle of repeated death and rebirth. Such an individual release—also known as *Moksha*—the liberation from Samsara (cycle of repeated death and rebirth) can be obtained through Moksha, which is the final aim in life of a human being and which can only be attained through *atma-jnama* (self-realisation).

However the departure of Kali Yuga would look like there is a possibility to escape, naturally through spirituality, moral behaviour, and through oneness with god. Whether Kali Yuga really functions as a warning against humanities' distance to god remains to be seen. In this case it would be possible to strategically utilise such a religious interpretation pattern. Salvation in this regard could include aspects of sustainability, and it could be one option to collaborate religious experts and communities.

4.4.5 Type E: Emissions, Climate Change, and Weather Changes

With this type, for the first time 'climate change' enters the stage. Respondents 13 and 14 do mention it explicitly, and they see human emissions as main drivers of a changing climate that in turn leads to perceived changes in weather patterns.

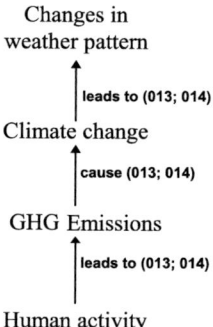

Figure 9: Type E: Emissions, Climate Change, and Weather Changes

---

[47] The prophets of the Old Testament often also predicted the 'end of the world' as a result of human sinfulness, but the social meaning of their prophecy was a call for re-moralisation of the world out of the free and substantially bettered will.

Table 10: Social characteristics of type E

Interview Code		13	14
Informants Age		81	45
Sex		Male	Female
GHG Emissions		1.44	4.99
Household income		350,000	2,220,000
Consumer Durables	Color TV	1	0
	LCD	1	1
	Refrigerator	1	1
	Fans	7	6
	Coolers	2	1
	ACs	2	3
	Cars	1	3
	Motorbikes / Scooters	0	0
Education		B. Sc. Mechanical Engineering	Bachelor of Science
Religion		Hindu	Hindu

Type E representatives are not explicit about the concrete nature of the human activities they mention, but we assume it would have been possible to let them enumerate some of them 'correctly'. However, as we confined ourselves to mapping what has been explicitly stated, this is the picture. From their social background the two representatives of this type belong to the middle and upper classes in India; MGI (2007) would classify them as 'seekers' and 'globals' respectively. This is also underlined by the household equipment structure and— to a lesser degree though—by personal GHG emissions. Respondent 13 is slightly above the Hyderabad average, while respondent 14 tops it more than three times (yet not the highest value in the sample!). Respondent 81 is an old, upper-cast male, a former high-level engineer from a family that can trace back its intellectual distinctiveness and bureaucratic professions to the days of the British Empire. While he can be regarded as a member of the 'old middle class', respondent 14, a woman working for a social NGO with a social science background, can be seen as a member of the 'new middle class' (cf. Fernandes 2007, 2009). Both have heard about climate change through the mass media, and (in case of respondent 13) due to professional interactions. The two did largely talk about issues of energy security and the future of the energy system. Respondent 14 had also heard about scientific discussions and remaining uncertainties with respect to the attribution of observed climate change to anthropogenic forcing. She nevertheless tended towards the attribution to humans.

From the perspective of an improved climate change discourse in Hyderabad, type E represents a rather simple, but scientifically more or less 'correct' representation of climate change. It would be necessary to reinforce people like our two respondents in their representation, possibly enriching it (see the next types), stabilising them against exaggerated climate 'skepticism' in the mass media, and try to use them as multipliers in their respective social groups and, if possible, for other social groups too.

4.4.6 Type F: Climate change due to ozone depletion

Two respondents did, in addition to climate change, bring ozone depletion into play, and they went further down the cause-effect chain of ozone depletion in order to explain climate change.

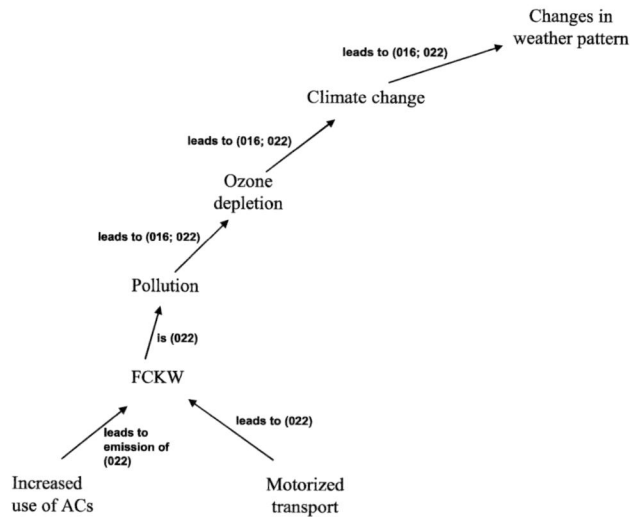

Figure 10: Type F: Climate change due to ozone depletion

Table 11: Social characteristics of type F

Interview Code		16	22
Informants Age		34	36
Sex		Female	Male
GHG Emissions		1.58	5.26
Household income		> 1,000,000	1,700,000
Consumer Durables	Color TV	3	1
	LCD	4	0
	Refrigerator	4	1
	Fans	23	5
	Coolers	1	1
	ACs	10	1
	Cars	9	1
	Motorbikes / Scooters	0	1
Education		MA Computer Science	BA Engineering
Religion		Hindu	Hindu

From a climate science point of view, the conceptualisation of ozone depletion as the major cause of climate change is incorrect. CFCs do play a role in both processes, but they drive climate change directly, not via the ozone layer depletion. The latter again is not a driver of climate change. Obviously, two loosely coupled environmental problems have been enmeshed here into one causal chain.

From a social science point of view, this confusion is understandable: ozone layer depletion and climate change both (1) are caused by humans, (2) play somewhere 'up there' in the atmosphere, (3) need political and everyday life fixing, and (4) less car driving seems to help against both. Respondent 22 mentioned motorised transport and increase in air conditioning as the main drivers of the whole chain, while respondent 16 did start with pollution as the common denominator for 'harming substances' that then trigger ozone depletion.

Both respondents belong to the upper and upper middle class, and their household equipment is above average—especially in the case of respondent 16 (23 fans, 9 cars in the household). Respondent 22 did travel a lot via airplane, leading to high per capita emissions. One thus cannot rule out the possibility that both people did perceive themselves as part of the problem (pollution or human activities leading to it).

Both respondents do have a high educational level (MA, BA), which most probably made them 'prone' to the adoption (and unfortunately: the mixture) of scientific arguments on climate change and ozone depletion—most probably from the mass media.

In order to communicate with adherents of type F, one would have to correct for the enmeshing of two human induced environmental problems. It is this type—and the representatives behind it—that can rather easily be engaged in a science-lay people dialogue that most scientists understand by 'scientific literacy'. The elements of their representation as well as their educational/professional background are highly science-oriented, and it should be easy to correct their mental model of climate change. At the same time it would be necessary and helpful to engage in a dialogue about possible solutions, and it is clear that their own lifestyle would then have to be addressed as well.

4.4.7 Type G: Multiple causation with ozone

Two respondents did enrich the cognitive map by two elements: first a more detailed causation pattern for ozone depletion, which then leads to climate change; and second they introduced deforestation into the picture.

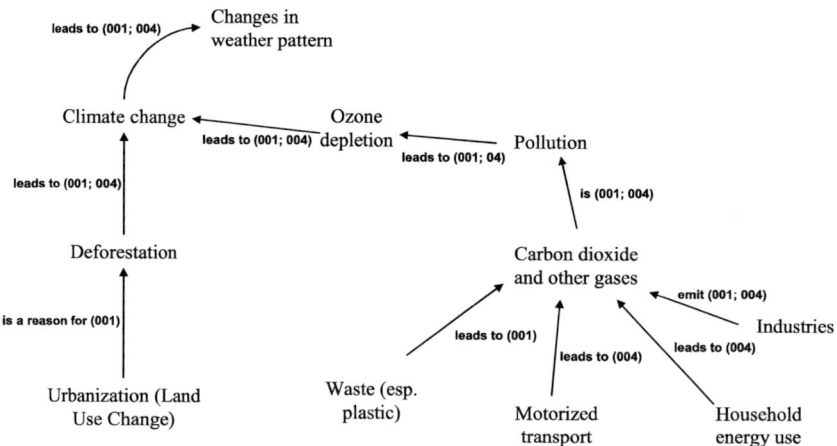

Figure 11: Type G: Multiple causation with ozone

Table 12: Social characteristics of type G

Interview Code		1	4
Informants Age		25	37
Sex		Male	Male
GHG Emissions [c]		3.57	10.72
Household income		100,000	360,000
Consumer Durables	Color TV	1	2
	LCD	0	1
	Refrigerator	0	1
	Fans	1	6
	Coolers	0	1
	ACs	0	2
	Cars	0	1
	Motorbikes/ Scooters	0	2
Education		B. Sc. Computer Engineering	Graduate in Public Personal Management and Sociology
Religion		Muslim	Hindu

[c] Overall yearly personal GHG emissions in tonnes CO_2 equivalents

From a scientific point of view, all arguments against type F (ozone → climate change) have to be repeated here. However, this type is characterised by a more detailed assessment of the causes of ozone depletion (industry, transport, household energy use, waste), and for the first time carbon dioxide has entered the stage. It thus seems as if type G does not (only) enmesh two problems, but rather *confuses* them, i.e. takes the correct causal antecedents for anthropogenic climate change and simply mixes this with the 'wrong' consequence, i.e. ozone depletion.

Deforestation due to land use changes (here mainly in an urbanisation context) is also a new element, but correctly offers a new element not to be found before.

The two respondents belong to the (lower) middle class. The very high per capita emissions of respondent No. 4 are due to a high professional travel budget, encompassing car, train, and airplane.

Again, a productive discourse on climate change on a mostly scientific basis could be possible with this group, as with group F. One could even engage in a solution discourse more intensely, as the causal pattern does address many human activities. And the explicit mentioning of deforestation (land use changes) offers a clear 'hook' to focus on issues of urban green and blue space, as well as on adaptation to climate change.

4.4.8 Type H: Multiple causation without ozone

Three respondents of our sample did adhere to a multiple causation model, in which climate change is responsible for weather changes, and many human activities (including deforestation) do cause climate change. Different to types F and G, ozone depletion does not play a causal role in explaining climate change. One can thus consider this type as basically 'correct' from a scientific point of view, although the ultimate causes at the human front end are more enumerations of various activities, a to some degree less comprehensive than in type G.

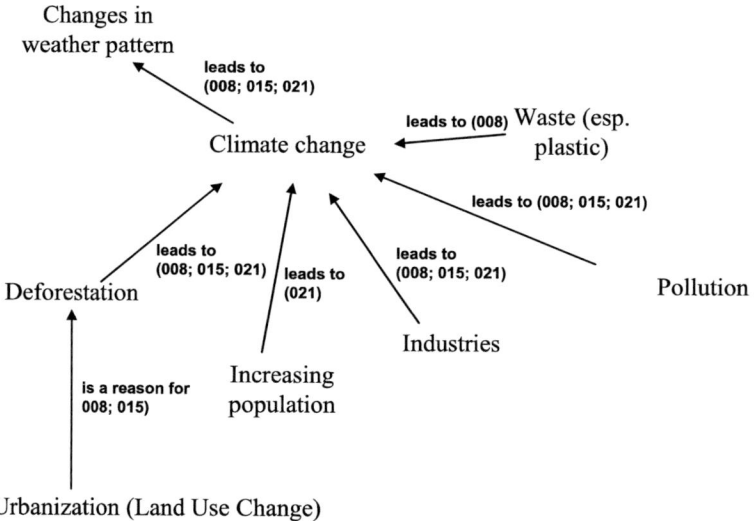

Figure 12: Type H: Multiple causation without ozone

As a new element, population growth comes up, while traffic is let out as a driver of climate change. Also, emissions of GHG do not explicitly occur. Nevertheless, as land use changes due to urbanisation processes and subsequent deforestation are mentioned, and as ozone depletion is missing, this type is very 'good' in terms of scientific results.

People from this group belong to the (upper) middle class, but their household equipment as well as their rather heterogeneous GHG emissions indicates some variance—which also means some leverage in terms of de-coupling modern lifestyles from carbon emissions. The lack of scientific elements—in part of 'misplaced' ones, such as ozone, in part of 'correct' ones, such as CO_2—can be interpreted as a positive sign, in the sense that these people are more interested in the human causation structure than in the physical mechanism that translates emissions into temperatures (and weather changes). This offers ample opportunities for a more 'humanistic' dialogue on climate change.

Table 13: Social characteristics of type H

Interview Code		21	8	15
Informants Age		34	37	58
Sex		Male	Female	Female
GHG Emissions		3.86	0.63	8.23
Household income		1,000,000	324,000	> 1,000,000
Consumer Durables	Color TV	1	1	2
	LCD	0	0	0
	Refrigerator	1	1	2
	Fans	4	3	10
	Coolers	1	0	1
	ACs	0	0	2
	Cars	1	0	3
	Motorbikes/ Scooters	1	0	1
Education		MBA	Bachelor in Commerce	Bachelor in Homoeopathy
Religion		Hindu	Hindu	Hindu

4.5 Solutions

We briefly touched the issue of solutions to climate change with those who did mention it. As we regard adaptation and mitigation as the two elements of the solution space, it is interesting to see that most respondents did immediately mention mitigation options, while adaptation has been widely left out. In Table 14 is a list of the topics mentioned.

In general, and not very surprisingly, the scope and the depths of solutions to climate change presented in the interviews is more or less directly coupled to the educational background of the respondent (dominant influence), and in part from his or her professional background (secondary influence). The higher educated members of our sample were able to reflect very clearly about their knowledge base, and where they themselves feel that it is lacking relevant information. This not only holds for the 'mechanism' of climate change and the certainty of its scientific basis. One respondent (No. 14) does explicitly mention the public debates about the 'failures' of IPCC in 2010 ('Glaciergate'), and Dr. Pachauri's role in particular. It does also hold for the solution space. New and cleaner technologies are mentioned, e.g. cleaner cars (either more fuel efficient, or based on renewable energy sources) or solar power energy systems. Many respondents see that government incentives and regulations have to change, because the current institutional structure is perceived as being inefficient. But our respondents did not simply blame the government. Many stressed the responsibility of individuals, especially of individuals with a lifestyle that contributes to pollution and unsustainable resource use. The rapid growth of the city and the lack of the public institutions to cope with it has been stressed (No. 22).

Table 14: Mitigation options

Category		No. of respondents
Solutions for mitigation of climate change (n=10)	Planting more trees	4
	People should start to think: "Starts from the people	3
	Better technologies (e.g. solar car)	3
	Strict control of old vehicles and polluting industries	2
	Using less plastic	2
	Education to the people	2
	Ban plastic	1
	Stop wasting resources	1
	Simple household energy saving measures	1
	Offsetting of air travel	1
	Change in diet (go organic)	1
	Make it mandatory for office buildings (e.g. in SEZs) use renewable energy	1
	Follow old cultures	1
	Responsibility of the industrialised countries	1
	Population control	1

Hardly any respondent was able to connect climate change with the impacts on people (was already the result of our expert interviews), let alone adaptation. The question was: What do you think will be the effects from climate change and what could be the solution to cope with these effects? Those respondents who actually replied to this question gave very vague answers, and in some cases these answers were rather related to mitigation options. A typical answer: "I guess it starts that we maybe need to look for other sources of energy, e.g. solar energy needs to be tapped, or nuclear energy. I am sure that definitely something has to be done, but I don't really know."

This uncertainty, if not even ignorance about the solution space increases the further down the social status ladder, especially the educational ladder we come. Although people from the lowest stratum of our sample feel a clear need for change (see the Kali Yug complex; but mostly with respect to environmental pollution in general, not with climate change), they were not able to specify concrete technological or institutional solutions. Instead, they seem to insist on a moral change, especially of those that have engaged in doubt- or harmful behaviours. It will be a challenge to reach out to these parts of Hyderabad in order to communicate the necessities of a low carbon lifestyle in the city. While they are well able to understand the moral need for it, they can hardly perceive the link to concrete measures at the level of technologies and institutions.

4.6 Environmental Pollution/ Ozone Depletion

In general it became obvious that people feel more comfortable within a topic that they can easily perceive with their senses. Therefore, environmental pollution is a good bridge to let people talk about problems that are indirectly linked to climate change like air pollution due to motorised traffic or indirect use of energy with regard to the perceivable characteristics of a throwaway society that slowly develops in the urban context in India. Most of the respondents refer to air pollution, many talked about waste and the use of plastic. Some respondents desire more strict rules to be enforced to control air pollution (discard old vehicles) and the use of plastic (many desired to ban plastic).

If we frame climate change causes with direct perceivable environmental pollution it will be much easier to derive mitigation options with significant co-benefits that are also perceived by the people as such. That means, e.g. that we don't have to tell them to switch to public transport, because it would be better for the climate; it would be easier to transport a message that directly impacts them, namely air pollution. It sounds very simple, but a context of nearly 40 % of illiteracy (whole India), at least it should not be taken for granted that people understand the links between the use of cars, air pollution, and health. How would they be able to understand the link to climate change?

Nevertheless, the high incidence of ozone depletion as a typical 'modern' way of (mis-) representing climate change does also indicate that one must not rely on the co-benefits from talking climate change in a pollution (abatement) framework. Large parts of the

Indian population are educated enough, embedded in globalised mass media communication, and engaged in emission-intensive lifestyles, so that following the pathway of a correction of the ozone model would also be promising.

5 Conclusions

In this report, we have dealt with the social representation of climate change by everyday social actors in Hyderabad (India). We understand social representations as 'ontologically' basic concepts, as they refer to the subjective and communicative (inter-subjective) dimension of real changes in atmospheric parameters together with their immediate drivers and underlying causes, as well as their natural and social impacts. Weather and climate cannot 'intervene' in society unless they become socially represented in one or the other way. And societies cannot intentionally act upon perceived or anticipated climate change if they do not represent it, e.g. by picking it up in everyday or professional communication.

We have conceptualised social representation as the semantics of a social discourse, i.e. as the explication of the meaning (content) in a particular form of knowledge on a special topic—here climate change and related issues like weather changes.

As we have seen in Chapter 1, based on the PIK work on downscaling of global climate models, climate change will most probably affect Hyderabad's climate. In particular, we expect higher average temperatures, more heat waves, and a changed seasonal precipitation regime with both a risk of flooding and of lacking rains. These changes bear the risk of adversely affecting specific urban functions, some urban impact hotspots, as well as particularly vulnerable social groups. In order to prepare for unavoidable climate change (adaptation), but also in order to reduce the urban greenhouse gas emissions as part of a locally accepted global responsibility and, hopefully, a strategy of a more sustainable development of the city, there is a clear need to translate scientific results on climate change into the urban society. A local climate discourse needs to be established—or reinforced.

We have seen that the science of climate change has been consolidated in the last years, and that it has made its way to the mass media, which we could show also for English speaking daily newspapers in India. Telugu speaking daily newspapers, however, are less explicit about climate change. If weather extremes are mentioned—which is the case—then climate change is hardly presented as a possible cause. This is a crucial finding, and it pre-determines the social representation of climate change to a large degree, at least for the (exclusive) readers of Telugu newspapers.

The Indian climate discourse is more complex (heterogeneous) than the European or American one, as we find almost all elements of these discourse, plus other elements that are unique to either a developing country or India in particular (especially social justice, actual impacts like heavy floods, and to some degree religious interpretations). Global comparisons reveal a rather lacking state of climate knowledge of Indian citizens. On the other hand, weather related news receive a high attention and offer a potential for linking them to climate change.

Our sample of 16 interviews is small and not representative, but informative and gives a first approximation of the social representation of climate change in Hyderabad. We found eight major types of representing climate change, including its nexus with weather changes and human causation. No weather changes and not mentioning climate change has been counted in here for systematic reasons. We find that both the pollution model (climate change is pollution, or driven by pollution) as well as the ozone model (climate change is driven by ozone depletion) do occur in the Hyderabad context, not only the pollution model, as one would expect for a developing country. We take this as a combined effect of the high level of education in India, a globalised mass media discourse, and the growing contribution of urban lifestyles in India to GHG emissions.

Despite of some misrepresentations of climate change, the *potential* for grasping both its mechanisms and the social drivers behind it is rather high in Hyderabad. One major reason for that is that the city is home to many members of the 'old middle class', mainly due to its function as a state capital, with many (leading) public administration people living there. On the other hand, Hyderabad has experienced more recently the expansion of the 'new middle class', mainly in IT and other modern service industries. Members of both 'fractions' of the Indian middle class are quite understandable about the emerging issue of climate change, they are aware of local environmental problems, and they see the link to the causes and drivers in a rapidly growing city. Whether or not this level of general environmental awareness (cf. Lange et al. 2009) does really translate into a new urban 'environmentality' (Agrawal 2005) remains to be seen for Hyderabad. Some studies from other metropolises in India give hope, especially if middle class organisations (such as Resident Welfare Organisations, RWA) do engage in urban governance and demand accountability (cf. Mooij and Lawa-Rewal 2010), while other examples raise scepticism whether the engagement for the common good really transcends individual / class egoism (Mawdsley 2009). However, the middle class representatives in our Hyderabad sample have been aware of the problems, and—at least in some initial form—even did reflect their own contribution to it, as well as the responsibility that flows from that reflection.

A major challenge will be to integrate the lower, less educated strata of the Hyderabad society in mitigation and adaptation projects. While the latter may resonate well with pressing needs in terms of (public) health, the former is much less coupled to a very morally determined worldview. Translating motivational capabilities, which are undoubtedly there, into concrete technological and institutional solutions in an understandable and convincing way will be important.

References

Adorno, T. W. and M. Horkheimer. 2002. *Dialectic of enlightenment*. Trans. Edmund Jephcott, Stanford: Stanford University Press.

Agrawal, B. 2005. *Environmentality: Technologies of government and the making of subjects*. Durham: Duke University Press.

Bapna, M. 2009. *Dispelling myths about India and climate change*. Washington DC.

Barbour, R. and U. Flick. 2009. "Doing focus groups." In *4: The Sage qualitative research kit*. edited by Uwe Flick. Reprinted., Los Angeles, Calif.: SAGE.

BBC World. 2007. "All countries need to take major steps on climate change: global poll." BBC World, London.

BBC World. 2010. "Africa talks climate: the public understanding of climate change in ten countries." BBC World/ British Council, London.

Beck, U. 2010a. "Klima des Wandels oder Wie wird die grüne Moderne möglich?" In *KlimaKulturen. Soziale Wirklichkeiten im Klimawandel.* edited by H. Welzer, H.-G. Soeffner and D. Giesecke: 33–48. Frankfurt am Main, New York, Campus.

Beck, U. 2010b "Climate for change, or how to create a green modernity?" *Theory, Culture & Society* 27 (2-3): 254–266.

Bhattacharya, S. et al. 2006. "Climate change and malaria in India" *Current Science* 90(3): 369–375.

Botzen, W.J.W., J.M. Gowdy and J.C.J.M. van den Bergh. 2008. "Cumulative CO_2 emissions: shifting international responsibilities for climate debt." *Climate Policy* 8: 569–576.

Boyce, T. and J. Lewis (eds.) 2009. "Climate change and the media." *Vol. 5: Global crises and the media.* New York: Peter Lang Publishing.

Boykoff, M. T. 2008. "Media and scientific communication: a case of climate change." *Geological Society* L. (Special Publication), London.

Boykoff, M. T. and J. M. Boykoff. 2004. "Balance as bias: global warming and the US prestige press." *Global Environmental Change* 14: 125–136.

Broad, K. and B. Orlove. 2007. "Channeling globality: The 1997–98 El Niño climate event in Peru." *American Ethnologist* 34(2): 285–302.

Buechler, S. and G. Devi. 2003. "Household Food Security and Waste water dependent Livelihood Activities along the Musi River in Andhra Pradesh, India." International Water Management Institute (IWMI).

Corfee-Morlot, J., M. Maslin and J. Burgess. 2007. "Global warming in the public sphere" *Philosophical Transactions of the Royal Society of London, Series A Mathematical, Physical and Engineering Sciences* 365(1860): 2741–2776.

Das, S., D. Mukhopadhyay and S. Pohit. 2005. "Mitigating Carbon Emission through Economic Instruments: An Indian Perspective." NCAER Working Paper 050001, National Council of Applied Economic Research. www.ncaer.org/Downloads/WorkingPapers/WP96.pdf [12-05-09].

Dumont, L. 2009. *Homo Hierarchicus. The Caste System and Its Implications.* Oxford etc.: Oxford University Press. [French edition 1979]

Edwards, P.N. 2001. "Representing the Global Atmosphere: Computer Models, Data, and Knowledge about Climate Change." In *Changing the Atmosphere. Expert Knowledge and Environmental Governance.* edited by C. A. Miller and P. N. Edwards: 31–65. Cambridge/London: The MIT Press.

Ernst, A. 2010. "Individuelles Umweltverhalten – Probleme, Chancen, Vielfalt." In *KlimaKulturen. Soziale Wirklichkeiten im Klimawandel.* edited by H. Welzer, H.-G. Soeffner and D. Giesecke: 128–143. Frankfurt am Main, New York, Campus.

Fernandes, L. 2007. *India's New Middle Class: Democratic Politics in an Era of Economic Reform* New Delhi: Oxford University Press.

Fernandes, L. 2009. "The Political Economy of Lifestyle: Consumption, India's New Middle Classes and State-Led Development." In: Lange and Meier (2009): 219-236.

Fleming, J. R. 1998. *Historical perspectives on climate change* New York: Oxford University Press.

Flick, U. 2006. *An introduction to qualitative research.* 3. ed., London: SAGE.

Giddens, A. 2009. *The politics of climate change.* Cambridge, Malden MA: Polity Press.

GoI (Government of India). 2008. *National Action Plan on Climate Change.* Delhi.

Greenpeace India. 2007. "Hiding Behind the Poor." A Report by Greenpeace India on Climate Injustice. Greenpeace India.

Gupta, D. (Ed.) 2010. *Social Stratification.* Oxford etc.: Oxford University Press [First edition 1981].

Habermas, J. 1981. *Theorie des kommunikativen Handelns.* Frankfurt am Main: Suhrkamp.

Harley, T. A. 2003. "Nice weather for the time of year: the British obsession with the weather." In *Weather, climate, culture*. edited by S. Strauss: 103–120. Oxford, Berg.

Homan, J. 2003. "The social construction of natural disaster: Egypt and the UK." In *Natural disasters and development in a globalizing world*. edited by M. Pelling: 141–157. London, Routledge.

HSBC. 2007. "HSBC climate confidence index 2007." HSBC.

Hubacek, K., D. Guan and A. Barua. 2007. "Changing lifestyles and consumption patterns in developing countries: A scenario analysis for China and India" *Futures* 39(9): 1084–1096.

IEA. 2006. *CO_2 Emissions from Fuel Combustion 1971–2004*. Paris.

Indraganti, M. 2010 "Using the adaptive model of thermal comfort for obtaining indoor neutral temperature: Findings from a field study in Hyderabad, India." *Building and Environment* 45(3): 519–536.

IPCC (Intergovernmental Panel on Climate Change). 2007. *Climate Change 2007: Synthesis Report*. IPCC, Geneva.

Jovchelovitch, S. 2001. "Social representations, public life and social construction" http://eprints.lse.ac.uk/2649 [11-06-10].

Kovats, S. and R. Akhtar. 2008. "Climate, climate change and human health in Asian cities." *Environ. Urban.* 20(1): 165–175.

Kuckartz, U. 2010. "Nicht hier, nicht jetzt, nicht ich – Über die symbolische Bearbeitung eines ernsten Problems." In *KlimaKulturen. Soziale Wirklichkeiten im Klimawandel*. edited by H. Welzer, H.-G. Soeffner and D. Giesecke: 144–160. Frankfurt am Main, New York, Campus.

Kumar, K. S., K. N. Tiwari and M. K. Jha. 2009. "Design and technology for greenhouse cooling in tropical and subtropical regions: a review." *Energy and Buildings* 41(12): 1269–1275.

Lamnek, S. 2005. *Qualitative Sozialforschung*. Lehrbuch 4., vollst. überarb. Aufl., Weinheim, Basel: Beltz.

Lange, H. and Meier, L. (eds.). 2009. *The New Middle Classes. Globalizing Lifestyle, Consumerism and Environmental Concern*. Dordrecht etc.: Springer.

Leiserowitz, A., E. Maibach and C. Roser-Renouf. 2010. "Climate change in the American mind: Americans' global warming beliefs and attitudes in January 2010." Yale University / George Mason, New Haven.

Lorenzoni, I. and N. Pidgeon. 2006. "Public views on climate change: European and USA perspectives." *Climatic Change* 77: 73–95.

Lüdeke, M.K.B. and M. Budde. 2009. "Evaluating climate change scenarios–From AOGCMs to Hyderabad." Project Report No. 1/2009, Potsdam/Berlin: Sustainable Hyderabad Project.

Mawdsley, E. 2009. "'Environmentality' in the Neoliberal City: Attitudes, Governance and Social Justice." In: Lange and Meier (2009): 237–251.

Mayring, P. 2002. *Einführung in die qualitative Sozialforschung.* Weinheim; Basel: Beltz.

McMichael, A. J. et al. 2004. "Global climate change." In *Comparative quantification of health risks. Global and regional burden of diseases attributable to selected major risk factors.* edited by M. Ezzati et al.: 1543–1649. Geneva.

MGI (McKinsey Global Institute). 2007. *The 'Bird of Gold': The Rise of India's Consumer Market.* San Francisco: McKinsey Global Institute.

Mooij, J. and S. T. Lama-Rewal. 2010. "Class in Metropolitan India: The Rise of the Middle Class." In: Ruet et al. (2010): 81–104.

Moscovici, S. 1961. *La psychanalyse, son image et son public.* Paris: P U F.

Moscovici, S. 1984. "The phenomenon of social representations." In *Social representations.* edited by R. M. Farr and S. Moscovici: 3–70. Cambridge, Cambridge University Press.

Moscovici, S. 1988. "Notes towards a description of social representations." *European Journal of Social Psychology.* 18(3): 211–250.

Müller, S. 2009. *Probleme der Dialektik heute.* Wiesbaden: VS Verlag für Sozialwissenschaften.

Mukhopadhyay, P. 2010. "India: When Urban Lifestyle Meets Climate Change." In *CITIES: steering towards sustainability.* edited by P. Jacquet, R. K. Pachauri and L. Tubiana: 69–79. New Delhi, TERI.

Nakićenović, N. and R. Swart (eds.). 2000. "Special Report on Emissions Scenarios. A Special Report of Working Group III of the Intergovernmental Panel on Climate Change." Cambridge, UK, New York: Cambridge University Press.

Nelson, L. E. 2000. *Purifying the earthly body of god: religion and ecology in Hindu India.* New Delhi: D.K. Printworld.

Nielsen, A. C. 2006. "Global Consumer Confidence & Opinions Survey." ACNielsen.

Nissel, H. 2009. "Contesting urban space: Megacities and globalization in India." *Geographische Rundschau (International Edition)* 5(1): 40–46.

O'Brien, K. et al. 2004. "Mapping vulnerability to multiple stressors: climate change and globalization in India." *Global Environmental Change* 14(4): 303–313.

O'Neill, S. J. and M. Hulme. 2009. "An iconic approach for representing climate change." *Global Environmental Change* 19(4): 402–410.

Orlove, B. 2005. "Human adaptation to climate change: a review of three historical cases and some general perspectives." *Environmental Science & Policy* 8(6): 589–600.

Oxfam America. 2004. "Weathering the Storm: lessons on risk reduction from Cuba."

Parikh, J. K. and K. Parikh. 2002. *Climate Change: India's Perceptions, Positions, Policies and Possibilities.* Paris: OECD.

Patt, A. 2001. "Understanding uncertainty: forecasting seasonal climate for farmers in Zimbabwe." *Risk, Decision and Policy* 6(2): 105–119.

Pinney, C. 1999. "On living in the kal(i)yug: notes from Nagda, Madhya Pradesh." *Contributions to Indian Sociology* 33(1-2): 77–106.

Ponting, C. 1993. *A green history of the world. The environment and the collapse of great civilizations.* New York, NY: Penguin Books.

Rahmstorf, S. 2010. "Das IPCC in der Medienkritik." *KomPass Newsletter* 11: 2–6.

Rahmstorf, S. and H. J. Schellnhuber. 2007. *Der Klimawandel.* München: Beck.

Ramesh, R. 2009. "India to reduce carbon intensity by 24 % by 2020." The Guardian 2.12.2009.

Reusswig, F. 2010. "Klimawandel und Gesellschaft. Vom Katastrophen- zum Gestaltungsdiskurs im Horizont der postkarbonen Gesellschaft." In *Der Klimawandel. Sozialwissenschaftliche Perspektiven.* edited by M. Voss: 75–97. Wiesbaden, Verlag für Sozialwissenschaften.

Reusswig, F., K. Gerlinger and O. Edenhofer. 2004. "Lebensstile und globaler Energieverbrauch. Analyse und Strategieansätze zu einer nachhaltigen Energiestruktur." Potsdam Institute for Climate Impact Research (PIK) (PIK Report 90), Potsdam.

Reusswig, F., L. Meyer-Ohlendorf and U. Anders. 2009. "Partners for a low-carbon Hyderabad - A stakeholder analysis with respect to lifestyle dynamics and climate change." Berlin/Potsdam: Sustainable Hyderabad Project, Background Study.

Reusswig, F., L. Meyer-Ohlendorf, D. Reckien, M.L.B. Lüdeke, S. Hofmann and O. Kit. 2009. "Is a low emission Hyderabad possible? Mapping urban transformations in a climate change scenario process." Paper presented at the IGU 2009 Conference *Emerging Urban Transformations: Multilayered Cities and Urban Systems*. Hyderabad, AP, India.

Reusswig, F., L. Meyer-Ohlendorf, U. Anders and A. Otto. 2009. "Climate change discourse in India: an analysis of press articles." Sustainable Hyderabad Project (Additional Study), Berlin, Potsdam.

Revi, A. 2008. "Climate change risk: an adaptation and mitigation agenda for Indian cities." *Environment and Urbanization* 20(1): 207–229.

Robertson, R. 1992. *Globalization social theory and global culture.* London: SAGE.

Roy, J. and S. Pal. 2009. "Lifestyles and climate change: link awaiting activation." *Current Opinion in Environmental Sustainability 2009* 1: 192–200.

Ruet, J. and S. T. Lama-Rewal (Eds.). 2009. *Governing India's Metropolises.* London/ New York/ New Delhi: Routledge.

Sahlins, M. D. 1976. *Culture and practical reason.* Chicago: University of Chicago Press.

Satterthwaite, D. et al. 2007. *Adapting to climate change in urban areas: the possibilities and constraints in low and middle income nations.* IIED (Human Settlements Discussion Paper 1), London.

Semenza, J. C. et al. 2008. "Public perception and behavior change in relationship to hot weather and air pollution." *Environmental Research* 107(3): 401–411.

Sen, A. 1981. *Poverty and Famines. An Essay on Entitlement and Deprivation.* Oxford and New York: Oxford University Press.

Shukla, P. R. et al. 2003. "Development and climate: an assessment for India." Indian Institute of Management (Report), Ahmedabad.

Sivak, M. 2009. "Potential energy demand for cooling in the 50 largest metropolitan areas of the world: implications for developing countries." *Energy Policy* 37(4): 1382–1384.

Skodvin, T. 2000. *Structure and Agency in the Scientific Diplomacy of Climate Change.* Dordrecht; Boston; London: Kluwer Academic Publishers.

Smith, N., J. Garrett and V. Vardhan. 2007. "Food and nutrition in Hyderabad. Current knowledge and priorities for action in an urban setting." Humboldt University (Research report 1), Berlin.

Strauss, A. L. 1987. *Qualitative analysis for social scientists.* Cambridge: Cambridge University Press.

Vairavamoorthy, K., S. D. Gorantiwar and A. Pathirana. 2008. "Managing urban water supplies in developing countries–Climate change and water scarcity scenarios." *Physics and Chemistry of the Earth* 33: 330–339.

van Rooijen, D. J., H. Turral and T. W. Biggs. 2005. "Sponge city: Water balance mega-city water use and waste water use in Hyderabad, India." *Irrigation and Drainage* 54: 581–591.

Vedwan, N. and R. E. Rhoades. 2001. "Climate change in the Western Himalayas of India: a study of local perception and response." *Climate Research* 19: 109–117.

Wagner, C. 2010. "India: a difficult partner in international climate policy." In *International climate policy: priorities of key negotiating parties.* edited by S. Dröge: 67–73. Berlin.

Walker, G. and D. King. 2008. *The hot topic: what we can do about global warming.* Orlando: Harcourt.

Weart, S. R. 2003. *The discovery of global warming, New histories of science, technology, and medicine.* Cambridge, MA: Harvard Univ. Press.

Weber, M. 1946. "The social psychology of the world religions." In *From Max Weber: Essays in Sociology.* edited by H. H. Gerth and C. Wright Mills: 267–301. New York, Oxford University Press.

Wisner, B. 2001. "Lessons from Cuba? Hurricane Michelle." November, 2001; http://online.northumbria.ac.uk/geography_research/radix/cuba.html [10-04-09].

World Resources Institute. 2010. "Climate Analysis Indicators Tool (CAIT) Version 7.0." Washington DC.

WWF. 2009. "Developing an engagement strategy for earth hour - India." Draft final report, New Delhi.

Yearley, S. 1994. "Social movements and environmental change." In *Social theory and the global environment.* edited by M. R. Redclift and T. Benton: 150–168. London , New York, Routledge.

Young, S., L. Balluz and J. Malilay. 2004. "Natural and technologic hazardous material releases during and after natural disasters: a review." *Science of the Total Environment* 322: 3–20.